Lecture Notes in Statistics

ctd. on inside back cover

Lecture Notes in Statistics

Edited by D. Brillinger, S. Fienberg, J. Gani,
J. Hartigan, and K. Krickeberg

40

Hans Rudolf Lerche

Boundary Crossing of Brownian Motion

Its Relation to the Law of the Iterated Logarithm and to
Sequential Analysis

Springer-Verlag
Berlin Heidelberg GmbH

Author

Hans Rudolf Lerche
Institut für Angewandte Mathematik, Im Neuenheimer Feld 294
6900 Heidelberg, Federal Republic of Germany

Mathematics Subject Classification (1980): 60 J 65, 62 L 10, 60 G 40, 60 F 10,
62 C 10, 58 G 11

ISBN 978-0-387-96433-1   ISBN 978-1-4615-6569-7 (eBook)
DOI 10.1007/978-1-4615-6569-7

Originally published by Springer-Verlag Berlin Heidelberg New York in 1986

2147/3140-543210

PREFACE

This is a research report about my work on sequential statistics during 1980 - 1984. Two themes are treated which are closely related to each other and to the law of the iterated logarithm:

I)   curved boundary first passage distributions of Brownian motion,

II) optimal properties of sequential tests with parabolic and nearly parabolic boundaries.

In the first chapter I discuss the tangent approximation for Brownian motion as a global approximation device. This is an extension of Strassen's approach to the law of the iterated logarithm which connects results of fluctuation theory of Brownian motion with classical methods of sequential statistics. In the second chapter I make use of these connections and derive optimal properties of tests of power one and repeated significance tests for the simplest model of sequential statistics, the Brownian motion with unknown drift. To both topics there underlies an asymptotic approach which is closely linked to large deviation theory: the stopping boundaries recede to infinity. This is a well-known approach in sequential statistics which is extensively discussed in Siegmund's recent book "Sequential Analysis". This approach also leads to some new insights about the law of the iterated logarithm (LIL). Although the LIL has been studied for nearly seventy years the belief is still common that it applies only for large sample sizes which can never be observed in practice. One of the goals of this study is to correct this belief somewhat by putting the LIL in (statistical) contexts where it gets a meaning for finite sample sizes.

This work has some overlap with Siegmund's book, although many of the approaches and results are different. Curved boundary crossing distributions is a joint theme but Siegmund emphazises likelihood-ratio arguments. Those are also used in Chapter II of this report, to study optimality in sequential testing, a theme which is essentially excluded from Siegmund's book.

I tried to keep the presentation of the subject as simple as possible, to present the basic relations between the different concepts in the simplest setting. This explains also the restriction to Brownian motion.

For statisticians the information might be of interest that Chapter II, the more statistical part of this work, can be read without knowledge of the contents of Chapter I. Only the introduction is a prerequisite.

Most of this work I did as a member of the Sonderforschungsbereich 123 at the University of Heidelberg and some as a visitor of the Mathematical Sciences Research Institute at Berkeley.

Finally I thank numerous friends and colleagues inside and outside of Germany for their interest and kind support.

Heidelberg, 13.7.1986   H.R. Lerche

CONTENTS

Introduction

Let W(t) denote the standard Brownian motion. Khintchine's law of the
iterated logarithm states that almost surely

$$(1) \qquad \limsup_{t \to \infty} \frac{|W(t)|}{\sqrt{2t \log \log t}} = 1 .$$

As Kolmogorov and Hartman-Wintner have shown, this law extends to partial
sums of independent identically distributed random variables with zero
mean and finite variance.

While the law of large numbers and the central limit theorem were in-
dispensible tools of mathematical statistics from the beginning, the law
of the iterated logarithm, which describes phenomena in the intermediate
domain of the two laws, for a long time seemed to have no statistical
significance at all. This view has changed somewhat in the last twenty
years since Neyman (1969), (1971) praised the new developments of sequential statistics connected with the tests of power one. Meanwhile,
proceding from the generalizations of the law of the iterated logarithm
due to Kolmogorov-Petrovski-Erdös and Strassen (cf. Strassen (1967)),
useful approximations of curved boundary first passage distributions
were derived, at first by Jennen-Lerche (1981) and then by Dinges
(1982), Ferebee (1982), (1983), Jennen (1985) and Klein (1986). Independently Cuzick (1981) found similar formulas by a different approach.

In the first chapter we develop certain aspects of these approximations,
mainly emphazising the connections to other methods and to the classical
fluctuation theory of Brownian motion.

In the second chapter we study the optimal properties of sequential
tests with parabolic and nearly parabolic boundaries (tests of power
one) when there is no indifference zone in the parameter space. This
research relates the work on sequential tests with (nearly) parabolic
boundaries due to Robbins, Siegmund et al. to the theory of optimal
sequential testing developed by Wald-Wolfowitz, Chernoff, Lorden et al..
The chapters are linked by the fact that information about curved boundary first passage distribution is needed to determine the optimal pro-

cedures. For a demonstration of this point see Section 2 of Chapter
II.

For simplicity we restrict our study to the model of Brownian motion.
This dispenses with the overshoot problems associated with random walks
that make the techniques more complicated. For more information on the
overshoot problems for curved boundary first passage distributions, see
the monographs of Siegmund (1985) and Woodroofe (1982).

Presumably Robbins in 1952 was the first who realized that the law of
the iterated logarithm (LIL) has some meaning for sequential statistics;
he noted that by the LIL the sequential version of the usual significance
test will, with probability one, eventually detect an effect even when
there is none. This negative observation started the study of sequential
tests with parabolic and nearly parabolic boundaries. Independently
of each other Darling-Robbins (1967) and Farrell (1964) found that the
LIL enables one to construct sequential procedures with discontinuous
power functions: the tests of power one. Their operating characteristics
were studied in detail by Darling-Robbins (1968), Robbins-Siegmund
(1970), (1973) and others. For an early survey see Robbins (1970). As
Neyman (1971) pointed out, Barnard had much earlier used a one-sided
sequential likelihood-ratio test as a control procedure having power
one in some range of the parameter space.

To explain the concept of tests of power one more thoroughly, let us
consider Brownian motion $W(t)$ with unknown drift θ. Let P_θ denote the
associated measure. A level α-test of power one for testing $H_0 : \theta = 0$
versus $H_1 : \theta \neq 0$ is given by a stopping time T of Brownian motion
$W(t)$ which satisfies

$$(2) \qquad P_0(T < \infty) = \alpha,$$

$$(3) \qquad P_\theta(T < \infty) = 1 \quad \text{for } \theta \neq 0.$$

Stopping always means a decision in favor of "$\theta \neq 0$".

Examples are provided by stopping times

(4) $T = \inf\{t>0 \mid |W(t)| \geq \psi(t)\}$

where $\psi(t)$ satisfies (i) $\psi(t)>c_o>0$ for all t; (ii) $\psi(t)$ is an upper class function of Brownian motion at infinity (this means $P_o(W(t)<\psi(t)$ for all t large) = 1); and (iii) $\psi(t)=o(t)$. By the law of the iterated logarithm (1), (i) and (ii) imply (2) for a certain $0<\alpha<1$ and by the law of large numbers (iii) implies (3). A popular example which satisfies these conditions is given by

(5) $\psi(t) = \left[(t+r)(\log(\frac{t+r}{r})+2\log b)\right]^{1/2}$ for $t > 0$ with $b > 1$.

For this example (2) holds with $\alpha=b^{-1}$ as is shown in Robbins-Siegmund (1970) (see Theorem 2.1 of Chapter I). The method which Robbins-Siegmund used is that of weighted likelihood-functions. It is a classical technique of sequential statistics due to Ville and Wald (see e.g. p. 75 of Wald's book), which has been extended by Siegmund (1977), Lalley (1982) and others. The result of Robbins-Siegmund (see Theorem 2.1, Chapter I) yields the crossing probabilities for Brownian motion over boundaries which belong to the upper class at infinity.

There is another classical method which yields crossing distributions for upper class boundaries at zero: the method of images (see e.g. Levy (1965, p. 82)). It will be presented in Section 1 of Chapter I in a rather general form following some ideas of Daniels (1982). From this general method of images the tangent approximation and the associated techniques of proof can be explained in a natural way (see Section 3 of Chapter I). Surprisingly the method of images and the method of mixtures of likelihood functions turn out to be equivalent up to time inversion (see Section 2). Further information on this topic is contained in Siegmund's monograph.

Besides these two methods there are not many others which lead to curved boundary probabilities for different types of boundaries. One which is beyond the scope of this study has been used by Novikov (1981) to prove Theorem 5.2 of Chapter I. It combines the Girsanov-transformation with Laplace-transform techniques.

The oldest results about curved boundary first passage distributions
are presumably those for exact parabolas. They rely on special tricks
which only work for these boundaries. A nice presentation of this subject is given by Uchiyama (1980).

The main topic of Chapter I is the tangent approximation, which compared
to the method of images and the method of weighted likelihood functions
yields results for a much wider class of boundaries. It is closely
related to the following classical results about Brownian motion.

a) The Bachelier-Levy formula (see Levy (1965)). It states that the density of the first passage time of Brownian motion over the boundary
$\psi(t)=\Lambda+bt$ is given by

$$(6) \qquad p(t) = \frac{\Lambda}{t^{3/2}} \quad \phi(\frac{\psi(t)}{\sqrt{t}})$$

where $\phi(y) = \frac{1}{\sqrt{2\pi}} e^{-y^2/2t}$. For a proof see also example 1 of Chapter I,
Section 1.

b) The Kolmogorov-Petrovski-Erdös (KPE) test (c.f. Ito-McKean (1974, pp.
33-34, 161-164) and Petrovski (1935), especially the footnote on p. 414
for some historical facts). Let $\psi(t)$ denote a positive continuous function on $\mathbb{R}_+ = (0,\infty)$ such that $\psi(t)/t$ is monotone decreasing and
$\psi(t)/\sqrt{t}$ is monotone increasing. For the standard Brownian motion W let
$S=\sup\{t>0\,|\,W(t)\geq\psi(t)\}$ denote the last entrance time below ψ. According
to a well known zero-one law, S is finite or infinite with probability
one. Then

$$P(S<\infty) = 1 \quad \text{if and only if} \quad \int\limits_1^\infty \frac{\psi(t)}{t^{3/2}} \phi(\frac{\psi(t)}{\sqrt{t}})\,dt < \infty.$$

The KPE-test is a generalisation of the LIL; for the functions $\psi(t) =$
$(2(1+\varepsilon)t \log \log t)^{1/2}$, $S = \infty$ with probability one if and only if $\varepsilon \leq 0$.

By the time inversion transformation $y = x/t$, $s = 1/t$, standard Brownian
motion on $(0,\infty)$ starting at 0) is mapped into itself with inverted time-
scale. The KPE-test as well as the LIL transform in the same way. Ito-

McKean and Strassen used this fact to give a direct and intuitive proof of the KPE-test. Ito-McKean's proof of the "if" direction (p. 34 in the 1974 edition) is based on a direct geometric argument. (Their "only if"-part (p. 161) uses some diffusion theory and is not direct at all.) Strassen (1967) gave a direct geometric proof of the "only if" part by showing a sharper result: Let $\psi(t)$ be an upper class function at zero and assume that $\psi(t)/t^{\alpha}$ is monotone decreasing for some $0<\alpha<1$ and that $\psi(t)$ satisfies some smoothness conditions. Let

(7) $T = \inf\{t>0 \mid W(t) \geq \psi(t)\}$

denote the first exit time of the Brownian motion W over ψ and p the density of its distribution. Strassen showed that as $t \to 0$, $p(t)$ can be approximated by the first exit density at t of the tangent to the curve ψ at t. (We shall call this the tangent approximation). By the Bachelier-Levy formula (6), Strassen's result that the tangent approximation holds can be written

(8) $p(t) = \dfrac{\Lambda(t)}{t^{3/2}} \phi\left(\dfrac{\psi(t)}{\sqrt{t}}\right)(1+o(1))$ as $t \to 0$

with $\Lambda(t) = \psi(t) - t\psi'(t)$. By time inversion Strassen derived from (8) the approximate densities of the tail distribution of the last entrance time S for upper class functions at infinity.

For completeness it should be mentioned that independently of Strassen, Daniels (1974) derived heuristically the tangent approximation for U-shaped boundaries as a local approximation.

Our study will discuss the tangent approximation as a global approxima-tion device. The following result describes our approach (see Theorem 4.1 of Chapter I). Let $\{\psi_a ; a \in \mathbb{R}_+\}$ denote an indexed family of smooth boundaries. Let T_a denote their associated first exit times (defined by formula (7)) and p_a their densities. Let $t_1 > 0$. Under conditions similar to those of Strassen and under the assumption, $P(T_a < t_1) \to 0$ as $a \to \infty$, the tangent approximation holds:

$$(9) \qquad p_a(t) = \frac{\Lambda_a(t)}{t^{3/2}} \; \phi(\frac{\psi_a(t)}{\sqrt{t}}) \; (1+o(1))$$

uniformly on $(0,t_1)$ as $a \to \infty$.

Since this approximation is uniform, we get an approximation of the first passage probabilities by integration. Examples are

$$(10) \qquad \text{(i)} \qquad \psi_a(t) = \sqrt{a}\psi(t/a), \text{ with } \psi \text{ a fixed function,}$$

$$\text{(ii)} \qquad \psi_a(t) = \sqrt{2(r+at)}, \; r > 0.$$

$$\text{(iii)} \qquad \psi_a(t) = at^\alpha, \; 0<\alpha<\frac{1}{2}.$$

Example (i) shows that Strassen's result on the tangent approximation is covered by (9). Example (ii) was studied by Siegmund (1977), extending the method of weighted likelihood functions. He derived

$$(11) \qquad P(|W(t)| > \sqrt{2(r+at)} \quad \text{for some } 0<t<t_1)$$

$$= \sqrt{a} \; e^{-a} \; \frac{2}{\sqrt{\pi}} \int_{\sqrt{2/t_1}}^{\infty} e^{-r\theta^2/2} \; \frac{d\theta}{\theta} \; (1 + o(1))$$

as $a \to \infty^*$.

Since the tangent approximation (9) holds uniformly on compact intervals for example (ii), one gets the first passage probability by integration:

$$(12) \qquad P(T_a<t_1) = \sqrt{a} \; e^{-a} \; \frac{1}{2\sqrt{\pi}} \int_0^{t_1} \frac{1+2r/at}{\sqrt{1+r/at}} \; e^{-r/t} \; \frac{dt}{t} \; (1+o(1)).$$

---

*The idea to use the tangent approximation as a global approximation device arose once during several discussions with H. Dinges about the meaning of Siegmund's results from the viewpoint of classical fluctuation theory of Brownian motion.

The leading term of the right hand side of (11) is twice that of (12), which follows by a coordinate transformation. Since (12) gives the probability for the one-sided problem, the two results agree.

There are some advantages to approximating the first passage densities rather than the probabilities. We demonstrate this in the case of example (iii). It can be shown (see Jennen-Lerche (1981)) that for $0<\alpha<1$ the tangent approximation holds uniformly on compact intervals, e.g. $[t_0,t_1]$ with $1<t_0<t_1<\infty$. By integration of equation (9) one obtains

(13) $\qquad P(t_0<T_a<t_1) = \dfrac{2(1-\alpha)}{2\alpha-1}\left[\Phi(at_1^{\alpha-1/2}) - \Phi(at_0^{\alpha-1/2})\right](1+o(1)).$

for $0<\alpha<1$. Its right hand side is equal to

$$
= \begin{cases} \dfrac{2(1-\alpha)}{2\alpha-1}\left[1-\Phi(at_0^{\alpha-1/2})\right](1+o(1)) & \text{if } \alpha > \tfrac{1}{2} \\[2em] \dfrac{2(1-\alpha)}{1-2\alpha}\left[1-\Phi(at_1^{\alpha-1/2})\right](1+o(1)) & \text{if } \alpha < \tfrac{1}{2} \end{cases}
$$

as $a \to \infty$.

This result shows that the first passage probabilities depend strongly on the growth rates of the boundaries. The probabilities for the examples (ii) and (iii), given by (11) and (13), are also quite different. On the other hand the tangent approximation of the densities always has the same form.

In Chapter I we will discuss several aspects of the tangent approximation such as its range, higher order refinements and numerical accuracy. Jennen in her doctoral thesis refined the tangent approximation by giving a second order approximation from which she derived Siegmund's refinement of (11) (see Theorem 4.4 of Chapter I). For another refinement see Ferebee (1983). The accuracy of the approximations is discussed after Theorem 4.4.

A central question of this study deals with the range of the tangent approximation: on which intervals does the tangent approximation hold uniformly?

It turns out that for lower class functions at infinity the tangent approximation cannot hold on $\mathbb{R}_+$ since by the KPE-test.

$$\int_1^\infty \frac{\Lambda_a(t)}{t^{3/2}} \; \phi \; (\frac{\psi_a(t)}{\sqrt{t}}) dt = \infty .$$

(For example (ii), this follows directly from equation (11).) This observation leads us to some interesting phenomena related to the KPE-test which will be described in Section 5 in detail: the tangent approximation of the hazard rate for killing of the process at the boundary. As a consequence we can state a uniform approximation result for lower class functions over the whole time axis and derive necessary and sufficient conditions for the tangent approximation (Theorem 5.5).

In the Supplement to Chapter I it is shown that the tangent approximation is a formal saddlepoint approximation. Some explanations of the organization of Chapter I are given there too.

The results of Chapter I are useful for a variety of applications, For instance in survival analysis they can be used to calculate the coverage probabilities of confidence bands of the Kaplan-Meier estimator constructed with curved boundaries. They can also be used to calculate the operating characteristics of sequential procedures approximately. According to a result of Le Cam (1979), Brownian motion with unknown drift for many sequential situations turns out as the limiting statistical model.

In Chapter II we shall investigate some optimal properties of sequential tests with parabolic boundaries. The results are of some interest for medical statistics, since the repeated significance test is currently in wide use and has until now had only a heuristical foundation. Armitage (1975) propagated the test in his monograph as a natural procedure for clinical trials. Its operating characteristics were studied by several authors (McPherson-Armitage (1971), Siegmund (1977), Woodroofe-Takahashi (1982)). But very little has been known about its optimality properties.

Our results extend some parts of the theory of sequential testing from the case when there is an indifference zone in the parameter space to the case when there is none. This branch of sequential testing has started with the optimality result of Wald-Wolfowitz (1948) about simple hypotheses and was continued by the work of Schwarz (1962), Kiefer-Sacks (1963), Lorden (1967) and Pollak (1978). All these authors studied sequential testing and design problems for composite hypotheses with an indifference zone by taking a Bayes approach and assuming that there are fixed costs c>0 per unit observation length. Following the idea of Chernoff (1959), letting c→0, they derived the asymptotic behavior of the rejection region. (Here an intrinsic relation to the approach of Chapter I shows up. Letting c → 0 means that the rejection boundaries recede to infinity. Therefore c^{-1} corresponds to the index a of the boundary ψ_a as considered in Chapter I).

It turned out that, as in the Wald-Wolfowitz result, for 0-1 loss the asymptotically optimal stopping rules are of the type: stop if the posterior probability of the hypothesis or the alternative is too small. We call such rules "simple Bayes rules".

There is a basic idea behind our approach. Differently than many of the previous authors did, we let the observation costs depend on the underlying parameter, in a mathematically convenient way. Then our results show that for 0-1 loss the simple Bayes rules are optimal or nearly optimal when there is no indifference zone. The special choice of the costs is discussed below.

One of the few problems of sequential statistics for which the sampling costs have been formulated to depend on the underlying parameter is the Anscombe-problem (see Anscombe (1963)). There it is asked for an optimal sequential sampling plan of a clinical trial under the assumption that the "costs" are proportional to the treatment differences. Recent progress on this problem has been made by Chernoff-Petkau (1981) and Lai-Robbins-Siegmund (1983). The last paper contains an interesting application of the tangent approximation.

For simplicity (for instance to prevent overshoot problems) we consider in Chapter II the statistical model of Brownian motion W(t) with drift $\theta \in \mathbb{R}$.

A problem of testing sequentially whether the drift is different from zero is the following. Let F be a prior on $\mathbb{R}$ given by
$F = \gamma\delta_0 + (1-\gamma)\int\phi(\sqrt{r}\theta)\sqrt{r}d\theta$ with $0<\gamma<1$ and $\phi(x) = \frac{1}{\sqrt{2\pi}}e^{-x^2/2}$, consisting of a point mass at $\{\theta = 0\}$ and a smooth normal part on $\{\theta \neq 0\}$. Let the sampling costs be $c\theta^2$, with $c > 0$, for the observation of W per unit time when the underlying measure is P_θ. We assume a loss function which is equal to 1 if $\theta = 0$ and a decision is made in favour of "$\theta \neq 0$" and which is identically 0 if $\theta \neq 0$. A statistical test consists of a stopping time T of Brownian motion where stopping means a decision in favour of "$\theta \neq 0$". The Bayes risk for this problem is given by

(14) $\rho(T) = \gamma P_0(T<\infty) + (1-\gamma)c\int_{-\infty}^{\infty}\theta^2 E_\theta T\phi(\sqrt{r}\theta)\sqrt{r}d\theta.$

The objective is to find a stopping rule T_c^* which minimizes (14). For cost c sufficiently small the optimal stopping rule T_c^* is a test of power one. A similar problem has been studied by Pollak (1978) who assumed an indifference zone in the parameter space.

The cost "$c\theta^2$" may at first seem unusual. The factor θ^2 is the Kullback-Leibler information number, $E_\theta\log\frac{dP_{\theta,1}}{dP_{0,1}}$, which quantifies the separation of the measures P_0 and P_θ. Its meaning is clarified by the following consideration. Let us consider two testing problems with simple hypotheses:

1) $H_0 : \theta = 0$ versus $H_1 : \theta = \theta_1$

2) $H_0 : \theta = 0$ versus $H_1 : \theta = \theta_2$

with $\theta_i > 0$, $i = 1,2,$. Let t_i, $i = 1,2$ denote the sampling lengths. The level-α Neyman-Pearson tests for both problems have the same power if and only if $\theta_1^2 t_1 = \theta_2^2 t_2$. (This follows from the power function of a Neyman-Pearson test of level α: $\Phi(-c_\alpha + \theta\sqrt{t})$). Thus the factor θ^2 standardizes the sampling lengths in such a way that the embedded simple testing problems are of equal difficulty. Besides this statistical aspect there is a basic mathematical reason for this choice of the sampling costs. Since in our decision problem (14), an indifference zone

does not occur and since $E_oT = \infty$ (because of $P_o(T < \infty) < 1$) we have
$\lim_{\theta \to 0} E_\theta T = \infty$. More information about the singularity is provided by a
lemma of Darling-Robbins (1968). It states that for every stopping rule
with $P_o(T < \infty) < 1$

(15) $E_\theta T \geq 2b/\theta^2$ where $b = -\log P_o(T < \infty)$.

Equality in (15) holds for the stopping rules

(16) $T_\theta = \inf\{t > 0 | \frac{dP_{\theta,t}}{dP_{o,t}} \geq e^b\}.$

Here $\frac{dP_{\theta,t}}{dP_{o,t}}$ denotes the likelihood ratio (Radon-Nikodym derivative)
of P_θ with respect to P_o given the path $W(u)$, $0 \leq u \leq t$. It is given by

$$\frac{dP_{\theta,t}}{dP_{o,t}} = \exp(\theta W(t) - \frac{1}{2}\theta^2 t).$$

According to (15) the expected sample size $E_\theta T$ of a test of power one
considered as a function of θ has a pole at $\theta = 0$. The choice of "c" or
"$c|\theta|$" instead of "$c\theta^2$" would imply that tests of power one have an
infinite Bayes risk for the prior we have chosen, since

$$\int |\theta|^i E_\theta T \phi(\sqrt{r}\theta)\sqrt{r}d\theta = \infty \quad \text{for } i=0,1.$$

A precise description of the pole of $E_\theta T$ is given by Farrell (1964),
Robbins-Siegmund (1973) and Jennen-Lerche (1982). For the result of the
latter see Corollary 5.3 of Chapter I. The sampling costs "$c\theta^2$" remove
the nonintegrability of the singularity of $E_\theta T$ for a large class of
tests of power one, although $\lim_{\theta \to 0} \sup \theta^2 E_\theta T = \infty$ still holds. For instance
for all tests of power one defined by

$$T = \inf\{t > 0 | |W(t)| \geq \psi(t)\}$$

where the function $\psi(t)$ is concave and $\psi(t) = o(t^{2/3-\varepsilon})$ when $t \to \infty$ (with $\varepsilon > 0$ arbitrary small), the Bayes risk (14) is finite. This follows from the inequality $\theta E_\theta T \leq \psi(E_\theta T)$, which is a consequence of Wald's lemma and Jensen's inequality. Therefore by the choice of the sampling costs as "$c\theta^2$" the concept of Bayes tests of power one becomes an interesting topic to study.

The related problem for simple hypotheses can be solved easily. The Bayes risk is given by

$$(17) \qquad \rho(T) = \gamma P_o(T<\infty) + (1-\gamma)c\theta^2 E_\theta T$$

and the issue is to find the minimizing stopping rule. A direct application of (15) and (16) yields the optimal stopping time

$$(18) \qquad T_c^* = \inf\{t>0 \,|\, W(t) \geq \log a/\theta + \tfrac{1}{2}\theta t\}$$

with $a = \gamma(2(1-\gamma)c)^{-1}$ when $a > 1$ and $T_c^* = 0$ otherwise. The minimal Bayes risk is then given by

$$(19) \qquad \rho(T_c^*) = 2(1-\gamma)c[\log a + 1].$$

(For more details see the last part of the proof of Theorem 4.2).

Here the choice of the sampling costs leads to a solution which is independent of θ. This becomes obvious when one expresses T_c^* in another way. It can be rewritten as $T_c^* = \inf\{t>0 \,|\, \gamma(W(t),t) \leq \frac{2c}{1+2c}\}$ where

$$\gamma(x,t) = \frac{\gamma}{\gamma + (1-\gamma)\dfrac{dP_{\theta,t}}{dP_{o,t}}(x)}$$

denotes the posterior probability of the parameter "0" at (x,t) with respect to the prior $F = \gamma\delta_o + (1-\gamma)\delta_\theta$. Then T_c^* has the intuitive meaning "stop when the posterior probability of the hypothesis "0" is too small". This is just a simple Bayes rule or equivalently the one-sided sequential probability ratio test (16).

The study in Chapter II shows that simple Bayes rules which stop when the posterior mass of the hypothesis "$\theta = 0$" is too small, are also nearly optimal for the risk (14). These rules were already discussed by Cornfield (1966) on a heuristical basis.

Each simple Bayes rule is a stopping rule of the type (4) with a boundary equal to $\psi(t) = \pm((t+r)(\log(\frac{t+r}{r}) + 2 \log b))^{1/2}$ with $b = \frac{\gamma}{2(1-\gamma)c}$. For large t this boundary asymptotically grows like $(t \log t)^{1/2}$, which is faster than the limiting growth rate $(2t \log \log t)^{1/2}$ of the law of the iterated logarithm. As a consequence of our results the minimal Bayes risk can be approximated by that of simple Bayes rules within $o(c)$ when $c \to 0$, which can be calculated. For the precise statement see the Theorems 1.1, 4.2, 4.3 and 4.4 of Chapter II and their corollaries. The one-sided problem is discussed in Theorem 4.5. The main result was first derived heuristically with the tangent-approximation. For this approach see Section 2.

The second problem discussed in Chapter II is testing the sign of the drift θ of Brownian motion W(t). About it a substantial literature already exists (e.g. Chernoff (1961), (1972), Bather (1962)). The parameter sets of the hypothesis H_0 and the alternative H_1 are given by $\Theta_0 = \{\theta < 0\}$ and by $\Theta_1 = \{\theta > 0\}$. We assume 0-1 loss, the usual loss structure for testing. The observation costs are chosen again as $c\theta^2$ where c is a positive constant and θ is the drift of the observed Brownian motion. On the parameter space $\Theta_0 \cup \Theta_1$ we put the normal prior $G(d\theta) = \phi(\sqrt{r}(\theta - \mu))\sqrt{r}d\theta$ with $\phi(y) = \frac{1}{\sqrt{2\pi}} e^{-y^2/2}$. The Bayes risk for a decision procedure (T,δ), consisting of a stopping time T of W and a final decision rule δ, is given by

$$(20) \qquad \rho(T,\delta) = \int_{-\infty}^{0} (P_\theta\{H_0 \text{ rejected } (\delta)\}+c\theta^2 E_\theta T)G(d\theta)$$

$$+ \int_{0}^{\infty} (P_\theta\{H_1 \text{ rejected } (\delta)\}+c\theta^2 E_\theta T)G(d\theta).$$

Let $G_{x,t}$ denote the posterior distribution of θ given that the process $(W(s),s)$ has reached (x,t). It is equal to $G_{x,t} = N(\frac{x+r\mu}{t+r}, \frac{1}{t+r})$ where $N(\rho,\sigma^2)$ denotes the normal distribution with mean ρ and variance σ^2. For $\lambda > 0$ we define the simple Bayes rule $T_\lambda = \inf\{t>0 \mid \min_{i=0,1} G_{W(t),t}(\theta_i) \leq \Phi(-\lambda)\}$ where Φ denotes the standard normal distribution function. It can also be expressed as

$$T_\lambda = \inf\{t>0 \mid \frac{|W(t)+r\mu|}{\sqrt{t+r}} \geq \lambda\}.$$

Let $\delta*$ denote the final decision rule which rejects H_o if and only if $W(T)+r\mu > 0$. In Theorem 5.1 it is shown that for a certain $\lambda(c)$, the pair $(T_{\lambda(c)}, \delta*)$ minimizes the Bayes risk (20). In Section 5 it is also explained why the repeated significance test is the natural counterpart to Wald's sequential likelihood ratio test for composite hypotheses.

This study is organized as follows. The first two sections of Chapter I describe two general methods for the calculation of curved boundary first passage distributions. In Section 2 it is shown that they are equivalent up to time inversion. In Section 3 from the general method of images (described in Section 1) the tangent approximation is derived. The tangent approximation is discussed for receding boundaries in Section 4 and for a fixed boundary when the time tends to infinity in Section 5. Also necessary and sufficient conditions and refinements for the tangent approximation are given there. In the Supplement to Chapter I the connection to saddlepoint approximations is discussed.

In Chapter II the first section introduces Bayes tests of power one, and the second gives a heuristic derivation of the optimal boundaries using the tangent approximation. In Section 3 some ideas about a statistical law of the iterated logarithm are discussed. Section 4 gives proofs about the shape of Bayes tests of power one, and Section 5 states an optimal property for the repeated significance test. The plan of this chapter follows the historical development of the results.

Some results which are stated in this work are not, at least partially, due to the author. These exceptions are the Theorems 1.3 (Widder), 2.1 (Robbins-Siegmund), 2.2 (Widder, Robbins-Siegmund) and Theorem 5.2 (Novikov) of Chapter I.

In Chapter I the first part of Section 3 is based on joint work with
H.E. Daniels. The results of Section 4 are mainly those of Jennen-Lerche
(1981), although the proof of Theorem 4.1 is new. Theorem 4.4 is a
version of a result of Jennen (1985). The results of Chapter II are
contained in the papers of Lerche (1985), (1986).

CHAPTER I

CURVED BOUNDARY FIRST PASSAGE DISTRIBUTIONS
OF BROWNIAN MOTION

1. The general method of images for the diffusion equation

The method of images for the diffusion equation is a simple analytical way to calculate first exit distributions of Brownian motion over curved boundaries. The idea behind it can be described in the following way: one considers a variety of "sources" which is distributed over the positive space axis at time zero according to a positive measure F and an extra source with unit mass at zero. The sources are thought of as the starting positions of Brownian motions to which one attributes negative weights according to F and a single positive unit weight when starting at 0. The superposition of all these processes is being observed.

The set of all space-time points in which the processes with positive and negative weights absorb each other, can be represented as a function of time called $\psi(t)$. Let us consider now the Brownian motion starting at 0 with absorption at the boundary ψ. On the set $\{(x,t)\,|\,x \le \psi(t)\}$ the method of images yields the distribution of the part of Brownian motion which is not absorbed at ψ (Theorem 1.1).

Using this result a rather explicit formula for the density of the first exit distribution can be given (Theorem 1.2). Theorem 1.3 states that the method of images is most general in some sense. Several examples demonstrate the use of the method.

For a precise description let us assume that F is a positive, σ-finite measure with $\int_0^\infty \phi(\sqrt{\epsilon}\theta) F(d\theta) < \infty$ for all $\epsilon > 0$. Here $\phi(x) = \frac{1}{\sqrt{2\pi}} e^{-x^2/2}$. Let $a > 0$ and let

$$(1.1) \qquad h(x,t) = \frac{1}{\sqrt{t}} \phi(\frac{x}{\sqrt{t}}) - a^{-1} \int_0^\infty \frac{1}{\sqrt{t}} \phi(\frac{x-\theta}{\sqrt{t}}) F(d\theta).$$

The function h satisfies the diffusion equation $\partial_t h = \frac{1}{2}\partial_x^2 h$ on $\mathbb{R} \times \mathbb{R}_+$ where $\mathbb{R}_+ = (0,\infty)$. The differentiation under the integral of (1.1) is valid by an argument similar to that of Copson (1975, p. 253). Let $x = \psi(t)$ denote the unique solution of the implicit equation

$$(1.2) \qquad h(x,t) = 0.$$

It exists since (1.2) after division by $\frac{1}{\sqrt{t}}\phi(\frac{x}{\sqrt{t}})$ is equivalent to the implicit equation

(1.3) $f(\frac{x}{t},\frac{1}{t}) = a$

with $f(y,s) = \int_0^\infty \exp(\theta y - \frac{1}{2}\theta^2 s) F(d\theta)$, for which an unique solution exists.

Let $W(t)$ denote the standard Brownian motion with starting point at 0 at time 0 and let T denote the stopping time defined by $T=\inf\{t>0\,|\,W(t)\geq\psi(t)\}$, which obviously can be rewritten as

$$T=\inf\{t>0\,|\,h(W(t),t)=0\}$$

$$=\inf\{t>0\,|\,f(\frac{W(t)}{t},\,\frac{1}{t}) = a\}.$$

Theorem 1.1: *Let* $C=\{(x,t)\,|\,x\leq\psi(t)\}$. *On* C *holds*

(1.4) $P(T>t,\ W(t)\in dx)=h(x,t)\,dx$

(1.5) $P(T\leq t\,|\,W(t)=x)=1-h(x,t)/(\frac{1}{\sqrt{t}}\,\phi(\frac{x}{\sqrt{t}})) = a^{-1}f(\frac{x}{t},\frac{1}{t}).$

Statement (1.5) follows from (1.4) by conditioning, which means by division with $\frac{1}{\sqrt{t}}\,\phi(\frac{x}{\sqrt{t}})$.

The following proof of (1.4) uses analytical tools. By an uniqueness theorem for the solutions of a certain boundary value problem of the forward diffusion equation the solutions constructed by the method of images and the probabilistic one given by $P(T>t,W(t)\in dx)/dx$, coincide. A probabilistic proof of statement (1.4), which uses martingale arguments, is sketched in Section 2.

<u>Proof</u>: The function h satisfies the following equations:

(1.6) $\partial_t h = \frac{1}{2}\partial_x^2 h$ on C ,

$$h(\psi(t),t)=0 \quad \text{for all } t>0,$$

$$h(\cdot,0)=\delta_o \quad \text{on } (-\infty,\psi(+0)].$$

Here δ_o denotes the Dirac-measure at 0. We note that $\psi(+0)=\lim_{t\searrow 0}\psi(t)$ exists according to Lemma 1.2 below. Now we apply some uniqueness arguments at first for a classical situation.

Case 1: We assume that $\theta^*>0$ where $\theta^*=\inf\{y|F(0,y]>0\}$.

According to Lemma 1.1 and 1.2, $\psi(+0)>0$ and ψ is infinitely often continuously differentiable. Now a uniqueness theorem for the boundary value problem (1.6) (cf. Friedman (1964), Theorem 16) yields that h is uniquely determined. On the other hand there is also a natural probabilistic solution of the boundary value problem (1.6). Let

$$P(T>t,\ W(t)\in dx) = p(x,t)dx \quad .$$

$p(x,t)$ denotes the density in the space-time point (x,t) of the part of Brownian motion which is not absorbed by the boundary ψ up to time t. The function p also solves the boundary value problem (1.6). Thus by the uniqueness theorem, p=h on C and (1.4) follows[*].

Case 2: Now we drop the assumption "$\theta^*>0$" and allow $\theta^*=0$.

For this case a general uniqueness theorem seems to be unknown and we therefore have to proceed somewhat differently.

Beside the original setup with functions h,ψ and a stopping time T we consider related situations defined as follows. For $\alpha>0$ let

$$h_\alpha(x,t):=\frac{1}{\sqrt{t}}\ \phi(\frac{x}{\sqrt{t}})-a^{-1}\int_\alpha^\infty \frac{1}{\sqrt{t}}\ \phi(\frac{x-\theta}{\sqrt{t}})F(d\theta)$$

---

[*] The construction of Friedman also yields that $h(x,t)$ is equal to the Green's function $G_C(0,0,x,t)$ of the problem, evaluated at $(0,0)$ and (x,t).

and let $\psi_\alpha(t)$ denote the solution of $h_\alpha(\psi_\alpha(t),t)=0$. T_α denotes the corresponding stopping time. Let $p_\alpha(x,t)=P(T_\alpha>t,W(t)\epsilon dx)/dx$. This setup satisfies the assumptions of case 1 and therefore statement (1.4) holds for it:

$$h_\alpha(x,t) = p_\alpha(x,t).$$

Now let $\alpha\to0$. Obviously $h_\alpha(x,t)\searrow h(x,t)$ (converges decreasingly) and $\psi_\alpha(t)\searrow\psi(t)$. Therefore also $h(x,t)=\lim_{\alpha\to0} p_\alpha(x,t)$ holds. Thus statement (1.4) follows if

(1.7) $\lim_{\alpha\to0} p_\alpha(x,t)=P(T>t,W(t)\epsilon dx)/dx$ holds.

But (1.7) is equivalent to

(1.8) $\lim_{\alpha\to0} P(T_\alpha>t\mid W(t)=x)=P(T>t\mid W(t)=x).$

The last statement follows from Lemma 1.3, if $\psi_\alpha\to\psi$ uniformly on $[0,t]$ and ψ is concave, differentiable and belongs to the upper class at $t=0$. Since $\psi_\alpha(s)\searrow\psi(s)$ for all $s\le t$, the uniform convergence follows. Lemma 1.1 yields the concavity of ψ. The upper class property of ψ can be seen as follows. According to the work of Petrowski (1935, p. 388), the function $h(x,t)$ is an "Irregularitätsbarriere" for the space-time point $(0,0)$ and therefore $(0,0)$ in the classical sense is a non-regular point for the Dirichlet-problem of the forward diffusion equation. Thus according to Theorem 3.1 of Doob (1955), the point $(0,0)$ is also non-regular in the probabilistic sense, which means that $P(T>0)=1$. But this means that ψ is an upper class function at $t=0$.

□□□

Integration yields the following corollary.

Corollary 1.1:

(1.9) $P(T\le t)=1-\Phi(\frac{\psi(t)}{\sqrt{t}})+a^{-1}\int_0^\infty\Phi(\frac{\psi(t)-\theta}{\sqrt{t}})F(d\theta).$

The properties of the boundaries obtained by the method of images are described in the following lemmas. Several examples are given below Theorem 1.2.

Lemma 1.1: The boundary ψ has the following properties:
a) ψ is infinitely often continuously differentiable,
b) $\psi(t)/t$ is monotone decreasing,
c) ψ is concave.

Proof: Statement a) follows from the theorem about the implicit functions applied to $f(\frac{x}{t},\frac{1}{t})=\int\exp(\theta x/t-\frac{1}{2}\theta^2/t)F(d\theta)$ since by definition

$$(1.10) \quad f(\frac{\psi(t)}{t},\frac{1}{t}) = a \quad \text{holds} \quad .$$

From (1.10) also b) follows.

To prove statement c) we put $\eta(s)=\psi(t)/t$ and $s=1/t$. Since $\eta(s)$ satisfies the implicit equation $a=\int\exp(\theta\eta(s)-\frac{1}{2}\theta^2 s)F(d\theta)$, Hölder's inequality yields that $\eta(s)$ is concave and thus $\eta''<0$. To show that $\psi(t)=t\eta(t^{-1})$ is concave, it is sufficient that $\psi''(t)<0$. But $\psi''(t)=\eta''(t^{-1})/t^3<0$.

□□□

Lemma 1.2: Let $\theta*=\inf\{y|F(0,y]>0\}\geq 0$. Then $\lim_{t\to 0}\psi(t)=\theta*/2$.

Proof: We argue similarly as Robbins-Siegmund (1973, p. 100). At first we show by contradiction that

$$(1.11) \quad \liminf_{t\to 0}\psi(t) \geq \theta*/2 \quad .$$

For this we assume that for an $0<\epsilon<1$ $\liminf_{t\to 0}\psi(t)\leq(\theta*-\epsilon)/2$. Thus for a certain sequence $t_i\to 0$

$$0<a=\int_{\theta*}^{\infty} e^{\theta\psi(t_i)/t_i-\frac{1}{2}\theta^2/t_i}F(d\theta)$$

$$\leq\int_{\theta*}^{\infty} e^{\frac{1}{2}\theta(\theta*-\epsilon)/t_i-\frac{1}{2}\theta\theta*(t_i^{-1}-\epsilon)}e^{-\frac{\epsilon}{2}\theta^2}F(d\theta)$$

$$\leq \int_{\theta*}^{\infty} e^{- \frac{\varepsilon\theta}{2}(t_i^{-1}-\theta*)} \ e^{-\frac{\varepsilon}{2}\theta^2} \ F(d\theta) \to 0 \ .$$

This contradiction proves (1.11). Similarly we get $\lim \sup_{t\to 0} \psi(t) \leq \theta*/2$, which together with (1.11) yields the statement of the lemma.

□□□

Lemma 1.3: _Let_ $\{\psi_n, n \geq 1\}$, _denote a sequence of positive boundaries which_ _(as_ $n \to \infty$) _converge uniformly on_ $[0,t]$ _to a boundary_ ψ. _Let_ $\psi_n \geq \psi$ _on_ $[0,t]$ _for all n and let_ ψ _be a differentiable concave upper class func-_ _tion of Brownian motion at 0. Then for every_ $x < \psi(t)$

$$\lim_{n\to\infty} P(W(u) \geq \psi_n(u) \quad for \ some \ 0 \leq u < t | W(t)=x)$$

$$= \quad P(W(u) \geq \psi(u) \quad for \ some \ 0 \leq u < t | W(t)=x).$$

Proof: Let $T_n = \inf\{u>0 | W(u) \geq \psi_n(u)\}$ and T be defined similarly. By the assumptions for a given $\alpha>0$ we can choose a $\delta>0$ such that $P(T<\delta)+P(t-\delta \leq T<t)<\alpha$. Let $\varepsilon>0$ be so small that $2\varepsilon(\varepsilon+\psi(\delta)+(t-\delta)\psi'(\delta)-x)/(t-\delta) \leq \alpha$. There exists an n_o such that $\sup_{0 \leq v \leq t}(\psi_n(v)-\psi(v))<\varepsilon$ for all $n \geq n_o$. Then

$$P(0<T<t | W(t)=x)-P(0<T_n<t | W(t)=x)$$

$$=P(0<T<t, T_n>t | W(t)=x)$$

$$=\int_o^t P(T\in du)P(T_n>t | W(u)=\psi(u), W(t)=x)$$

$$\leq \alpha + \int_\delta^{t-\delta} P(T\in du)P(W(s)<\psi(s)+\varepsilon \quad for \ u \leq s \leq t | W(u)=\psi(u), W(t)=x).$$

By the concavity of ψ and a result about straight-line crossing proba-bilities (see Example 1 below)

$$P(W(s)<\psi(s)+\varepsilon \text{ for } u\leq s\leq t \mid W(u)=\psi(u),W(t)=x)$$

$$\leq P(W(v)<\varepsilon+\psi(u)+v\psi'(u),0<v<t-u \mid W(0)=\psi(u),W(t-u)=x)$$

$$=1-\exp(-2\varepsilon(\varepsilon+\psi(u)+(t-u)\psi'(u)-x)/(t-u)) .$$

Thus the above inequalities can be continued by

$$\leq\alpha+2\varepsilon(\varepsilon+\psi(\delta)+(t-\delta)\psi'(\delta)-x)/(t-\delta)$$

$$\leq 2\alpha \quad \text{for all } n\geq n_o .$$

□□□

Let p(t) denote the density of the distribution of T. It exists and is continuous since ψ is continuously differentiable (cf. Strassen (1967)). The following well known lemma states precisely what is clear intuitively; p can be computed from h by differentiation.

Lemma 1.4:

$$p(t) = -\frac{1}{2}\partial_x h(x,t)\Big|_{x=\psi(t)} .$$

Proof: By Theorem 1.1 we have

$$(1.12) \quad P(T>t) = \int_{-\infty}^{\psi(t)} h(y,t)dy .$$

Noting that $h(\psi(t),t)=0$, differentiation of (1.12) yields

$$-2p(t)=2 \frac{d}{dt} P(T>t)$$

$$=2 \frac{d}{dt}(\int_{-\infty}^{\psi(t)} h(y,t)dy)$$

$$= 2 \int_{-\infty}^{\psi(t)} \partial_t h(y,t) \, dy + 2\psi'(t) h(\psi(t),t)$$

$$= \int_{-\infty}^{\psi(t)} \partial_y^2 h(y,t) \, dy$$

$$= \partial_x h(x,t) \big|_{x=\psi(t)} \quad .$$

□□□

With the help of this lemma the density p can be calculated rather explicitly.

Theorem 1.2:

$$(1.13) \quad p(t) = \frac{E(\theta \mid (\psi(t),t))}{2t^{3/2}} \, \phi\left(\frac{\psi(t)}{\sqrt{t}}\right)$$

where

$$E(\theta \mid (\psi(t),t)) = \frac{\int_0^{\infty} \theta \phi\left(\frac{\theta - \psi(t)}{\sqrt{t}}\right) F(d\theta)}{\int_0^{\infty} \phi\left(\frac{\theta - \psi(t)}{\sqrt{t}}\right) F(d\theta)} \quad ,$$

Proof: By Lemma 1.4 and the definition of h, given by (1.1), we get

$$p(t) = -\frac{1}{2} \frac{\partial h(x,t)}{\partial x} \bigg|_{x=\psi(t)}$$

$$= \frac{\psi(t)}{2t^{3/2}} \, \phi\left(\frac{\psi(t)}{\sqrt{t}}\right) - a^{-1} \int_0^{\infty} \frac{\psi(t) - \theta}{2t^{3/2}} \, \phi\left(\frac{\psi(t) - \theta}{\sqrt{t}}\right) F(d\theta)$$

$$= \frac{1}{2t^{3/2}} \, \phi\left(\frac{\psi(t)}{\sqrt{t}}\right) \left[\psi(t) - a^{-1} \frac{\int_0^{\infty} (\psi(t) - \theta) \phi\left(\frac{\psi(t) - \theta}{\sqrt{t}}\right) F(d\theta)}{\phi\left(\frac{\psi(t)}{\sqrt{t}}\right)} \right]$$

$$= \frac{1}{2t^{3/2}} \; \phi(\frac{\psi(t)}{\sqrt{t}}) \left[\psi(t) - \frac{\int_0^\infty (\psi(t)-\theta) \phi(\frac{\psi(t)-\theta}{\sqrt{t}}) F(d\theta)}{\int_0^\infty \phi(\frac{\psi(t)-\theta}{\sqrt{t}}) F(d\theta)} \right]$$

$$= \frac{1}{2t^{3/2}} \; \phi(\frac{\psi(t)}{\sqrt{t}}) E(\theta | (\psi(t),t)) \; .$$

The step preceding the last one, follows from the definition of ψ since

$$\frac{1}{\sqrt{t}} \; \phi(\frac{\psi(t)}{\sqrt{t}}) \; = \; a^{-1} \int_0^\infty \frac{1}{\sqrt{t}} \; \phi(\frac{\psi(t)-\theta}{\sqrt{t}}) F(d\theta) \quad .$$

□□□

A similar result holds when we consider instead of Brownian motion without drift such one with drift. We denote by $p_\xi(t)$ the density of the distribution of T under the measure of Brownian motion with drift ξ. By the Cameron-Martin-Girsanov formula (see Liptser-Shiryayev (1977)) we have

(1.14) $p_\xi(t) = e^{\xi\psi(t) - \frac{1}{2}\xi^2 t} \; p_0(t)$.

Combining this formula with (1.13) yields the following result.

Corollary 1.2:

(1.15) $p_\xi(t) = \dfrac{E(\theta | (\psi(t),t))}{2t^{3/2}} \; \phi(\dfrac{\psi(t)-\xi t}{\sqrt{t}})$.

It is quite interesting to note that the factor $E(\theta|(\psi(t),t))$ does not change if the drift changes. The quantity $E(\theta|\psi(t),t)$ can be interpreted as the posterior expectation, with respect to the prior F, of the hitting point θ on the vertical axis at time 0 of the backward running Brownian motion. Its meaning in the formula (1.15) will become clearer in the next sections. Before we give several examples we note that similar results hold also for the two-sided boundaries, which can be constructed as follows. For a positive measure with mass on $(-\infty, +\infty)$

and with $F(\{0\})=0$, we define

$$h(x,t) = \frac{1}{\sqrt{t}} \, \phi(\frac{x}{\sqrt{t}}) - a^{-1} \int_{-\infty}^{\infty} \frac{1}{\sqrt{t}} \, \phi(\frac{x-\theta}{\sqrt{t}}) F(d\theta) \ .$$

There exist positive and negative valued functions ψ_+ and ψ_- with $\psi_- < \psi_+$ with the properties $h(\psi_+(t),t)=0$ and $h(\psi_-(t),t)=0$ for all $t<t_a$ with a certain $t_a \leq \infty$. When we define the stopping time T as

$$T = \inf\{0 < t \leq t_a \,|\, W(t) \notin (\psi_-(t), \psi_+(t))\},$$

then the analogous results to the preceeding ones hold.

Now we present several examples including also those with two-sided boundaries. For further informations see Daniels (1982).

Example 1: $F(d\theta) = \delta_{2\theta}$. Then

$$h(x,t) = \frac{1}{\sqrt{t}} \, \phi(\frac{x}{\sqrt{t}}) - a^{-1} \frac{1}{\sqrt{t}} \, \phi(\frac{x-2\theta}{\sqrt{t}}) \ ,$$

$$\psi(t) = \theta + bt \quad \text{with } b = \frac{\log a}{2\theta} \ .$$

The Theorems 1.1 and 1.2 yield

$$P(T \leq t \,|\, W(t) = x) = \exp(-\frac{2\theta}{t}(\theta+bt-x)) \quad \text{for } x \leq \psi(t),$$

$$P(T \leq t) = 1 - \Phi(\frac{bt+\theta}{\sqrt{t}}) + e^{-2\theta b} \phi(\frac{bt-\theta}{\sqrt{t}}) \ ,$$

$$p(t) = \frac{\theta}{t^{3/2}} \, \phi(\frac{\psi(t)}{\sqrt{t}}) \ , \text{ the well-known Bachelier-Levy formula.}$$

Example 2: $F(d\theta) = \alpha_1 \delta_\theta + \alpha_2 \delta_{2\theta}$ with $\alpha_1 + \alpha_2 = 1$.

$$h(x,t) = \frac{1}{\sqrt{t}} \, \phi(\frac{x}{\sqrt{t}}) - a^{-1}\left[\frac{\alpha_1}{\sqrt{t}} \, \phi(\frac{x-\theta}{\sqrt{t}}) + \frac{\alpha_2}{\sqrt{t}} \, \phi(\frac{x-2\theta}{\sqrt{t}})\right] \ .$$

The function ψ satisfies the equation

$$a = \alpha_1 \exp\left(-\frac{\theta^2}{2t} + \frac{\theta\psi(t)}{t}\right) + \alpha_2 \exp\left(-\frac{2\theta^2}{t} + \frac{2\theta\psi(t)}{t}\right) \ .$$

It can be expressed as

$$\psi(t) = \frac{1}{2}\theta - \frac{t}{\theta} \log\left\{\frac{1}{2}\frac{\alpha_1}{a} + \left(\frac{1}{4}\left(\frac{\alpha_1}{a}\right)^2 + \frac{\alpha_2}{a} e^{-\theta^2/t}\right)^{1/2}\right\}.$$

For $t \to 0$ holds $\psi(t) \sim \frac{1}{2}\theta + b_1 t$ and
for $t \to \infty$ holds $\psi(t) \sim b_2 t$ with

$$b_1 = \log(a/\alpha_1)/\theta \text{ and } b_2 = \log(a/\alpha_2)/(2\theta) \ .$$

$$P(T \le t \mid W(t) = x) = \exp\left(-\frac{\theta}{t}\left(\frac{\theta}{2} + b_1 t - x\right)\right) + \exp\left(-\frac{2\theta}{t}(\theta + b_2 t - x)\right)$$

$$\text{for } x \le \psi(t) \ .$$

$$p(t) = \frac{E(\theta \mid (\psi(t), t))}{t^{3/2}} \phi\left(\frac{\psi(t)}{\sqrt{t}}\right) \quad \text{with}$$

$$E(\theta \mid (\psi(t), t)) = \frac{\theta\alpha_1 \phi\left(\frac{\theta - \psi(t)}{\sqrt{t}}\right) + 2\theta\alpha_2 \phi\left(\frac{2\theta - \psi(t)}{\sqrt{t}}\right)}{\alpha_1 \phi\left(\frac{\theta - \psi(t)}{\sqrt{t}}\right) + \alpha_2 \phi\left(\frac{2\theta - \psi(t)}{\sqrt{t}}\right)} \ .$$

Example 3: $F(d\theta) = \frac{1}{2}\delta_{-\theta} + \frac{1}{2}\delta_\theta$, $\theta > 0$.

$$h(x,t) = \frac{1}{\sqrt{t}} \phi\left(\frac{x}{\sqrt{t}}\right) - a^{-1} \frac{1}{2\sqrt{t}}\left[\phi\left(\frac{x+\theta}{\sqrt{t}}\right) + \phi\left(\frac{x-\theta}{\sqrt{t}}\right)\right] \ ,$$

$$\psi_\pm(t) = \pm t/\theta \cosh^{-1}(a \exp(\theta^2/2t)).$$

These functions are only well defined for t with $1 < a \exp(\theta^2/2t)$. This is always the case for $a > 1$. For $a < 1$ the boundary is closed at

$t_a = \theta^2/(2 \log a^{-1})$. As Daniels (1982) pointed out, for a=1 the boundary is open and behaves for large t like a parabola

$\psi_{\pm}(t) \sim \pm (t + \frac{1}{6}\theta^2)^{1/2} + 0(t^{-1})$.

$$P(T \leq t \mid W(t) = x) = a^{-1} \exp(-\theta^2/2t) \cosh(\frac{x\theta}{t})$$

for all $|x| \leq \psi_{+}(t)$ and all a and t with $1 < a \exp(\theta^2/2t)$. Since

$$E(\theta \mid (\psi_{+}(t), t)) = \theta \frac{\sinh(\frac{\psi_{+}(t)\theta}{t})}{\cosh(\frac{\psi_{+}(t)\theta}{t})} \quad ,$$

$$p_{+}(t) = \frac{\theta}{2t^{3/2}} \frac{\sinh(\frac{\psi_{+}(t)\theta}{t})}{a \exp(\theta^2/2t)} \phi(\frac{\psi_{+}(t)}{\sqrt{t}}) \quad .$$

Example 4: $F(d\theta) = \frac{d\theta}{\sqrt{2\pi}}$ on $\mathbb{R}$.

$$h(x,t) = \frac{1}{\sqrt{t}} \phi(\frac{x}{\sqrt{t}}) - a^{-1}/\sqrt{2\pi} \quad .$$

$$\psi_{\pm}(t) = \pm\sqrt{t \log(\frac{a^2}{t})} \qquad \text{for } 0 < t \leq a^2.$$

$$P(T \leq t \mid W(t) = x) = \left[\frac{a}{\sqrt{t}} \phi(\frac{x}{\sqrt{t}})\right]^{-1} \qquad \text{for } |x| \leq \psi_{+}(t) \text{ and } t < a^2.$$

$$p_{+}(t) = \frac{\psi_{+}(t)}{2t^{3/2}} \phi(\frac{\psi_{+}(t)}{\sqrt{t}}) \quad .$$

Here $\quad E(\theta \mid (\psi_{+}(t), t)) = \psi_{+}(t)$.

Example 5: $F(d\theta) = e^{-\lambda\theta} \dfrac{d\theta}{\sqrt{2\pi}}$ on $\mathbb{R}$, $\lambda \in \mathbb{R}$

$$h(x,t) = \frac{1}{\sqrt{t}}\, \phi(\frac{x}{\sqrt{t}}) - a^{-1}\phi(\frac{x}{\sqrt{t}}) \exp(\frac{(x-\lambda t)^2}{2t}).$$

$$\psi_{\pm}(t) = \lambda t \pm \sqrt{t\,\log(\frac{a^2}{t})}.$$

$$E(\theta \mid (\psi_{\pm}(t),t)) = \psi_{\pm}(t) - \lambda t$$

$$p_{\pm}(t) = \frac{\psi_{\pm}(t) - \lambda t}{t^{3/2}}\, \phi(\frac{\psi_{\pm}(t)}{\sqrt{t}}) \ .$$

This result describes the boundary of Example 4 shifted by λt. It agrees with Corollary 1.2.

Example 6: $F(d\theta) = \sqrt{r}\, \phi(\sqrt{r}\theta)d\theta$ on $\mathbb{R}$.

$$h(x,t) = \frac{1}{\sqrt{t}}\, \phi(\frac{x}{\sqrt{t}}) - (\frac{rt}{rt+1})^{1/2} \exp(\frac{x^2}{2t(rt+1)} - \frac{x^2}{2t}) a^{-1}/\sqrt{2\pi} \ .$$

$$\psi_{\pm}(t) = \pm((rt+1)(\log a^2 + \log(\frac{rt+1}{rt})))^{1/2},$$

$$P(T \leq t \mid W(t) = x) = a^{-1}(\frac{rt}{rt+1})^{1/2} \exp(\frac{x^2}{2t(rt+1)}) \quad \text{for } |x| \leq \psi_{+}(t).$$

$$E(\theta \mid (\psi_{+}(t),t)) = \frac{\psi_{+}(t)}{rt+1} \ .$$

Thus $\quad p_{+}(t) = \dfrac{\psi_{+}(t)}{2(rt+1)t^{3/2}}\, \phi(\dfrac{\psi_{+}(t)}{\sqrt{t}}) \ .$

Example 7: $F(d\theta) = \dfrac{d\theta}{\sqrt{2\pi}}\, 1_{(0,\infty)}$.

$$h(x,t) = \frac{1}{\sqrt{t}}\, \phi(\frac{x}{\sqrt{t}}) - a^{-1}\phi(\frac{x}{\sqrt{t}})/\sqrt{2\pi} \ .$$

$\psi(t)$ is given as the solution of the implicit equation

$$\frac{x^2}{t} + 2 \log \Phi(\frac{x}{\sqrt{t}}) = \log(\frac{a^2}{t}).$$

Let $h(y) = y^2 + 2 \log \Phi(y) \sim y^2$ for $y \to \infty$. Then $\psi(t) = \sqrt{t} \; h^{-1}(\log(\frac{a^2}{t}))$. It holds $\psi(t) = \sqrt{t \log(1/t)}(1+o(1))$ for $t \to 0$ and $\psi(t) = -\frac{t}{a}(1+o(1))$ for $t \to \infty$. Let $\pi(y) = \phi(y)/\Phi(y)$. Then

$$P(T \le t \mid W(t) = x) = (a \; \pi(\frac{x}{\sqrt{t}}))^{-1} \; ,$$

$$E(\theta \mid (\psi(t), t)) = \psi(t) + \sqrt{t} \; \pi(\frac{\psi(t)}{\sqrt{t}}) \; ,$$

$$p(t) = (\frac{\psi(t)}{2t^{3/2}} + \frac{\pi(\frac{\psi(t)}{\sqrt{t}})}{2t}) \; \phi(\frac{\psi(t)}{\sqrt{t}}) \; .$$

Finally a remark on the generality of the approach just presented. According to the following Choquet-type representation theorem due to Widder (1944), the method of images uses all positive harmonic functions on $\mathbb{R} \times \mathbb{R}_+$ as will be explained below.

For further information on this type of theorem see Doob (1984, p. 290) and Robbins-Siegmund (1973).

Theorem 1.3: *Let u be a non-negative and continuous function on* $I = \mathbb{R} \times (0, \delta)$, $0 < \delta \le \infty$. *The following statements are equivalent:*

(i) *u satisfies the diffusion equation* $\partial_t u = \frac{1}{2} \partial_x^2 u$ *on I and*
$$\lim_{(x,t) \to (\xi, 0)} u(x,t) = 0 \text{ for all } \xi < 0.$$

(ii) *There exists a positive σ-finite measure F on* $[0, \infty)$ *such that u can be represented as*

$$u(x,t) = \int_0^\infty \frac{1}{\sqrt{t}} \; \phi(\frac{x-\theta}{\sqrt{t}}) F(d\theta) \; .$$

The meaning of the result for the method of images is the following. Let ψ denote a smooth upper class boundary at zero and let

$T=\inf\{t>0\,|\,W(t)\geq\psi(t)\}$. Let $p(x,t)=P(T>t,W(t)\in dx)/dx$. Then on $C=\{(x,t)\,|\,x<\psi(t)\}$, p can be expressed as $p(x,t)=\frac{1}{\sqrt{t}}\phi(\frac{x}{\sqrt{t}})-g(x,t)$ where g fulfills the diffusion equation on C. If g satisfies the diffusion equation in the whole plane $\mathbb{R}\times\mathbb{R}_+$ and is positive there, then by Theorem 1.3, p has the form (1.1), which shows that the method of images is a rather general approach.

2. The method of weighted likelihood functions

A probabilistic way to derive first exit distributions over straight lines and curved boundaries is presented in this section. It uses the fact that mixtures of likelihood functions are positive martingales. Although the basic result of this section is well known (cf. Robbins-Siegmund (1973)), its relation to other methods for computing exit distributions was left in the dark until recently. Surprisingly the connection between the approach described here and the general method of images, described in the preceding section, is basic and simple: both methods are equivalent up to time inversion. We shall develop this connection in detail. It will lead us to a simple martingale proof of Theorem 1.1 at the end of this section.

For simplicity we again consider the one-sided case. Let F be a positive σ-finite measure on $\mathbb{R}_+$. Let

$$f(x,t) = \int \exp\left(\theta x - \frac{1}{2}\theta^2 t\right) F(d\theta)$$

and let us assume that there exists a point (x_o, t_o) with $f(x_o, t_o) < \infty$. Then $f(x,t) < \infty$ for all $x \in \mathbb{R}$ and $t > t_o$. This is a consequence of the equation

$$f(x,t) = \left(\phi\left(\frac{x-x_o}{\sqrt{t-t_o}}\right)\right)^{-1} \int_o^\infty \phi\left(\sqrt{t-t_o}\left(\theta - \frac{x-x_o}{t-t_o}\right)\right) e^{\theta x_o - \frac{1}{2}\theta^2 t_o} F(d\theta).$$

Let $0 < a < \infty$ and let us denote the solution of the equation

$$f(x,t) = a$$

by $x = \eta_a(t)$ for $t \geq t_o$. This solution is uniquely determined for all $t \geq t_o$ and

(2.1) $f(x,t) < a$ if and only if $x < \eta_a(t)$

holds. Similar arguments as in Lemma 1.1 yield that η_a has the following properties for $t \geq t_o$: η_a is monotone increasing, concave and infinitely often continuously differentiable. For the standard Brownian motion $W(t)$ we define the stopping time

(2.2) $T = \inf\{t > t_o \mid f(W(t), t) \geq a\}$

if the infimum exists and $T = \infty$ otherwise. By (2.1) we get the equivalent representation

(2.3) $T = \inf\{t > t_o \mid W(t) \geq \eta_a(t)\}$.

The following result is due to Robbins-Siegmund (1970).

Theorem 2.1: Let $0 < f(x_o, t_o) \leq a < \infty$. *Then*

(2.4) $P(t_o \leq T < \infty \mid W(t_o) = x_o) = a^{-1} f(x_o, t_o)$.

Integration of (2.4) yields the following corollary.

Corollary 2.1:

(2.5) $P(t_o \leq T < \infty) = 1 - \phi\left(\dfrac{\eta(t_o)}{\sqrt{t_o}}\right) + a^{-1} \int_0^\infty \phi\left(\dfrac{\eta(t_o) - \theta t_o}{\sqrt{t_o}}\right) F(d\theta)$.

For the proof of Theorem 2.1 it is convenient to introduce some additional notation.

$P_\theta^{(x,t)}$ denotes the measure of Brownian motion W with drift θ which starts in x at time t. It is defined on the space $C[t, \infty)$ of continuous functions on $[t, \infty)$. F_s^t denotes the σ-algebra on $C[t, \infty)$ which is generated by $W(u)$, $t \leq u \leq s$. The restriction of the measure $P_\theta^{(x,t)}$ on F_s^t is denoted by $P_{\theta,s}^{(x,t)}$. This notation is also used for stopping times S instead of fixed times s. The measures $P_{\theta,s}^{(x,t)}$ and $P_{\theta',s}^{(x,t)}$ with $\theta \neq \theta'$ are absolutely continuous with respect to each other. By the Cameron-Martin-

Girsanov formula the Radon-Nikodym derivative (the likelihood function) on F_s^t is given by

$$(2.6) \qquad \frac{dP_{\theta,s}^{(x,t)}}{dP_{\theta',s}^{(x,t)}} = \exp\left((\theta-\theta')(W(s)-x)-\tfrac{1}{2}(\theta^2-\theta'^2)(s-t)\right)$$

The following proof uses a well known argument due to Ville and Wald.

Proof of Theorem 2.1: Let

$$F_{x_o,t_o}(d\theta) = \frac{dP_{\theta,t_o}^{(0,0)}}{dP_{o,t_o}^{(0,0)}}(x_o)F(d\theta) \Big/ \int \frac{dP_{\theta,t_o}^{(0,0)}}{dP_{o,t_o}^{(0,0)}}(x_o)F(d\theta).$$

Let $Q^{(x_o,t_o)}$ denote the probability measure
$Q^{(x_o,t_o)} = \int P_\theta^{(x_o,t_o)} F_{x_o,t_o}(d\theta)$. On the σ-algebra $F_t^{t_o}$ with $t>t_o$, $Q^{(x_o,t_o)}$ and $P_o^{(x_o,t_o)}$ are absolutely continuous with respect to each other. By (2.6) the Radon-Nikodym derivative is given by

$$(2.7) \qquad L_t = \frac{dQ_t^{(x_o,t_o)}}{dP_{o,t}^{(x_o,t_o)}} = \int \frac{dP_{\theta,t}^{(x_o,t_o)}}{dP_{o,t}^{(x_o,t_o)}} F_{x_o,t_o}(d\theta)$$

$$= \int \exp\left(\theta(W(t)-x_o)-\tfrac{1}{2}\theta^2(t-t_o)\right) F_{x_o,t_o}(d\theta)$$

$$= \frac{f(W(t),t)}{f(x_o,t_o)}.$$

Since L_t^{-1} is a positive $Q^{(x_o,t_o)}$-martingale the optional stopping theorem, (2.7) and (2.3) (the definition of T) yield the following equation:

$$(2.8) \qquad P_o^{(x_o,t_o)}(t_o \leq T \leq t_1) = \int_{\{t_o \leq T \leq t_1\}} L_{t_1}^{-1}\, dQ^{(x_o,t_o)}$$

$$= \int_{\{t_o \leq T \leq t_1\}} L_T^{-1} dQ^{(x_o, t_o)}$$

$$= f(x_o, t_o) \int_{\{t_o \leq T \leq t_1\}} f(W(T), T)^{-1} dQ^{(x_o, t_o)}$$

$$= a^{-1} f(x_o, t_o) Q^{(x_o, t_o)} \{t_o \leq T \leq t_1\}.$$

We show now that $\lim_{t_1 \to \infty} Q^{(x_o, t_o)} \{t_o \leq T \leq t_1\} = 1$, which together with (2.8)

yields the proof.

Let $\theta^* = \inf\{\theta | F(0, \theta) > 0\}$. It is sufficient to show that for all $\theta_o > \theta^*/2$

(2.9) $P_{\theta_o}^{(x_o, t_o)} (t_o \leq T < \infty) = 1.$

Let $\tilde{\theta} > 0$ and $\varepsilon > 0$ be chosen such that

$$\theta^* \leq \tilde{\theta} \leq \tilde{\theta} + \varepsilon < 2\theta_o \quad \text{with} \quad F([\tilde{\theta}, \tilde{\theta} + \varepsilon]) > 0.$$

$$f(W(t), t) = \int \exp(\theta W(t) - \tfrac{1}{2}\theta^2 t) F(d\theta)$$

$$\geq \min_{\theta \in [\tilde{\theta}, \tilde{\theta} + \varepsilon]} \exp(\theta W(t) - \tfrac{1}{2}\theta^2 t) \, F[\tilde{\theta}, \tilde{\theta} + \varepsilon].$$

The exponent of this expression for $t \to \infty$ grows under P_{θ_o} at least like

$\min_{\theta \in [\tilde{\theta}, \tilde{\theta} + \varepsilon]} \theta[\theta_o - \tfrac{1}{2}\theta] t$, which by the choice of $\tilde{\theta}$ tends to infinity. By the

definition of T this yields (2.9) and completes the proof.

□□□

Example 1: $F = \delta_{2\theta}$. Then $\eta_a(t) = b + \theta t$ with $b = \log a / 2\theta$. By Theorem 2.1

$$P(W(t) \geq b + \theta t \quad \text{for some } t > t_o | W(t_o) = x_o) = \exp(-2\theta(b + \theta t_o - x_o)),$$

and by Corollary 2.1

$$P(W(t) \geq b+\theta t \quad \text{for } t > t_o) = 1 - \phi(\frac{b+\theta t_o}{\sqrt{t_o}}) + e^{-2\theta b}\phi(\frac{b-\theta t_o}{\sqrt{t_o}}) \ .$$

Example 2: $F(d\theta) = \frac{d\theta}{\sqrt{2\pi}}$ on $\mathbb{R}$. Here $f(x,t) = t^{-1/2}\exp(x^2/2t)$ and $\eta_a(t) = \pm(t(\log t + \log a^2))^{1/2}$ which is defined for $t > a^{-2}$. Then by Theorem 2.1

$$P(t_o \leq T < \infty | W(t_o) = x_o) = a^{-1}t_o^{-1/2}\exp(x_o^2/2t_o) \quad \text{for } t_o > a^{-2}.$$

Example 3: $F(d\theta) = \sqrt{r}\phi(\sqrt{r}\theta)d\theta$ on $\mathbb{R}$. Then $f(x,t) = \sqrt{\frac{r}{t+r}}\exp(x^2/2(t+r))$ and $\eta_a(t) = \pm((t+r)(\log a^2 + \log(\frac{t+r}{r})))^{1/2}$. Theorem 2.1 yields

$$P(t_o \leq T < \infty | W(t_o) = x_o) = a^{-1}(\frac{r}{t_o+r})^{1/2}\exp(x^2/2(t_o+r)).$$

We note that this expression is equal to a^{-1} for $x_o = t_o = 0$.

Example 4: Let $\delta > 0$.

$$F(d\theta) = \delta d\theta/(\theta(\log 1/\theta)\ldots(\log_n (1/\theta))^{1+\delta}) \quad \text{for } 0 < \theta < 1/e_n.$$

Here $\log_2 x = \log(\log x)$, $e^2 = e^e$ etc.
Robbins-Siegmund (1970) have shown that for $n \geq 3$

$$\eta_a(t) = [2t(\log_2 t + \frac{3}{2}\log_3 t + \sum_{k=4}^{n} \log_k t + (1+\delta)\log_{n+1} t + \log(\frac{1}{2}a/\sqrt{\pi}) + o(1)] \quad \text{as } t \to \infty.$$

Here $P(T < \infty | W(0) \neq 0) = a^{-1}$.

Example 5: $F(d\theta) = \gamma \exp(-\alpha\theta^{-\beta})d\theta$ on $\mathbb{R}_+$, with $\alpha, \beta, \gamma > 0$ and $\alpha(\beta)$ and $\gamma(\beta)$ chosen such that F is a probability measure. According to Robbins-Siegmund (1970)

$$\eta_a(t) = t^{\frac{1+\beta}{2+\beta}} (1+o(1)) \quad \text{as } t \to \infty \quad \text{and}$$

$$P(T<\infty|W(0)=0) = a^{-1}.$$

Example 6: It is possible to develop the method described in this section further such that one can also calculate certain probabilities for bounded exit times. Using the martingale of Example 3, Siegmund (1977) has shown that

$$P(|W(t)| \geq \sqrt{2a(r+t)} \quad \text{for some } 0<t\leq at_1) =$$

$$= \sqrt{a} \; e^{-a} \int_{\sqrt{2/t_1}}^{\infty} e^{-r\theta^2/2} d\theta/\theta \; (1+o(1))$$

as $a \to \infty$.

Now we explain the relation between the method of weighted likelihood functions and the method of images. At first we repeat the definitions of the boundaries $\psi_a(s)$ and $\eta_a(t)$. According to (1.2) and (1.3) $\psi_a(s)$ is defined as the solution $y=\psi_a(s)$ of the implicit equation

$$(2.10) \quad f(\frac{y}{s}, \frac{1}{s}) = a \quad \text{with} \quad f(x,t) = \int_0^{\infty} \exp(\theta x - \frac{1}{2}\theta^2 t) F(d\theta).$$

According to (2.2) $\eta_a(t)$ is defined as the solution $x=\eta_a(t)$ of the implicit equation $f(x,t)=a$. The equations (2.2) and (2.10) together yield that the boundaries for $t=\frac{1}{s}$ satisfy the equations

$$(2.11) \quad \frac{\psi_a(s)}{s} = \eta_a(t) \quad \text{and} \quad \psi_a(s) = \frac{\eta_a(t)}{t}.$$

Now Theorem 1.1 states that for the Brownian Bridge W_0 which starts in 0 at time 0 and ends in $y_0 \leq \psi_a(s_0)$ at time s_0

$$(2.12) \quad \tilde{P}_{(0,0)}^{(y_0,s_0)}(W_0(s) \geq \psi_a(s) \quad \text{for some } 0<s\leq s_0) = a^{-1} f(\frac{y_0}{s_0}, \frac{1}{s_0}).$$

On the other hand Theorem 2.2 states that for the Brownian motion W starting in x_0 at time t_0 holds

$$(2.13) \quad P^{(x_0, t_0)} (W(t) \geq \eta_a(t) \quad \text{for some } t_0 \leq t < \infty) = a^{-1} f(x_0, t_0).$$

Now we note, that under the time-inversion transformation

$$(2.14) \quad x = \frac{y}{s}, \quad t = \frac{1}{s},$$

a Brownian Bridge with endpoints $(\theta, 0)$ and (y_0, s_0), is mapped into a Brownian motion with drift θ starting at $(\frac{y_0}{s_0}, \frac{1}{s_0})$ and running up to infinity. Therefore the formulas (2.12) and (2.13) are equivalent up to time inversion (this means the one formula follows from the other by applying the time inversion transformation (2.14) and vice versa). The reader is invited to check some of the examples of both sections on their correspondence.

These observations raise the further leading question: how are the methods themselves related?

The probabilistic meaning of the transformation (2.14) corresponds to the following analytical meaning which was discovered by Appell (1892). Under analysts the transformation (2.14) is also known as the Appell-transformation. Let u satisfy the forward diffusion equation $\partial_t u = \frac{1}{2} \partial_x^2 u$ in a region D_u of the space-time plane $\mathbb{R} \times \mathbb{R}_+$. Let $f(y,s) = u(\frac{y}{s}, \frac{1}{s}) / \sqrt{s} \phi(\frac{y}{\sqrt{s}})$. Then f satisfies the backward diffusion equation

$$\partial_s f + \frac{1}{2} \partial_y^2 f = 0 \quad \text{in } D_f = \{ (y,s) \mid (\frac{y}{s}, \frac{1}{s}) \in D_u \}$$

and vice versa. Thus the Appell-transformation establishes a duality relation between solutions of the forward and backward diffusion equations. We can use this fact to dualize Theorem 1.3, the representation theorem of Widder, to one for the harmonic functions of the backward diffusion equation. Here we give a version due to Robbins-Siegmund (1973) on $\mathbb{R}$.

Theorem 2.2: Let f *be a non-negative and continuous function on* $D = \mathbb{R} \times (\tau, \infty)$ *where* $0 \leq \tau < \infty$. *The following statements are equivalent:*

i) $\partial_s f + \frac{1}{2} \partial_y^2 f = 0$ *on* D.

ii) *There exists a positive* σ*-finite measure* F *such that*

$$f(x,t) = \int_0^\infty \exp(\theta x - \frac{1}{2}\theta^2 t) F(d\theta).$$

iii) $f(W(t),t)$ *is a positive martingale of the space-time Brownian motion on* D.

From the Theorems 1.3 and 2.2 we conclude that the method of images and the method of weighted likelihood functions use the different characterizing properties of the same objects, the harmonic functions for the forward and backward diffusion equations. We conclude further from the representation theorems that both methods use all the non-negative harmonic functions.

Both methods also have the same serious drawback. That is, only for few mixing measures the implicitly defined boundaries can be calculated explicitly. Robbins-Siegmund studied in several papers how the boundaries depend on the underlying mixing measure and constructed boundaries with the most typical growth rates (see p. 36 and 37).

The problem, which concave boundaries can be constructed from the two described methods, is to our knowledge not yet solved.

We close this section with a sketch of a probabilistic proof of Theorem 1.1 by dualizing the proof of Theorem 2.1 with the help of (2.14).

A probabilistic proof of Theorem 1.1, (1.5):

Let $\tilde{P}_{(\theta,0)}^{(y_0,s_0)}$ denote the measure of the Brownian Bridge process $W_0(u)$ with endpoints $(\theta,0)$ and (y_0,s_0). Let $\tilde{F}_s^{s_0}$ denote the σ-algebra which is generated from $W_0(u)$, $s \leq u \leq s_0$. The measures $\tilde{P}_{(\theta,0)}^{(y_0,s_0)}$ and $\tilde{P}_{(0,0)}^{(y_0,s_0)}$ are absolutely continuous with respect to each other on $\tilde{F}_s^{s_0}$ and their likelihood-ratio is given by

$$(2.15) \quad \frac{d\tilde{P}^{(y_0,s_0)}_{(\theta,0)}}{d\tilde{P}^{(y_0,s_0)}_{(0,0)}} \Big|_{\tilde{F}^{s_0}_s} = \exp(\theta(\frac{W_0(s)}{s} - \frac{y_0}{s_0}) - \frac{1}{2}\theta^2(\frac{1}{s} - \frac{1}{s_0})).$$

This can be seen either by direct calculation or by applying the time inversion transformation to the formula (2.6). Let

$$F_{y_0,s_0}(d\theta) = \frac{\phi(\frac{\theta-y_0}{\sqrt{s_0}})F(d\theta)}{\int\phi(\frac{\theta-y_0}{\sqrt{s_0}})F(d\theta)}$$

$$= \frac{\exp(\theta\frac{y_0}{s_0} - \frac{1}{2}\theta^2/s_0)F(d\theta)}{\int\exp(\theta\frac{y_0}{s_0} - \frac{1}{2}\theta^2/s_0)F(d\theta)}.$$

Let $\tilde{Q}^{(y_0,s_0)} = \int\tilde{P}^{(y_0,s_0)}_{(\theta,0)}F_{y_0,s_0}(d\theta)$. Then by (2.15)

$$(2.16) \quad \tilde{L}_s = \frac{d\tilde{Q}^{(y_0,s_0)}}{d\tilde{P}^{(y_0,s_0)}_{(0,0)}} \Big|_{\tilde{F}^{s_0}_s} = f(\frac{W_0(s)}{s}, \frac{1}{s})/f(\frac{y_0}{s_0}, \frac{1}{s_0}) \quad \text{holds}$$

with $f(z,v) = \int\exp(\theta z - \frac{1}{2}\theta^2 v)F(d\theta)$. Let $\sigma = \sup\{0 < s \leq s_0 | W_0(s) \geq \psi_a(s)\}$ if the supremum exists and $\sigma = 0$ otherwise. σ is a stopping time with respect to the σ-algebras $\tilde{F}^{s_0}_s$, $s \leq s_0$ and can also be expressed as $\sigma = \sup\{0 < s \leq s_0 | \tilde{L}_s \geq a\}$. Then by the optional stopping theorem and similar arguments as for (2.4), it follows by (2.16)

$$\tilde{P}^{(y_0,s_0)}_{(0,0)}\{W_0(s) \geq \psi_a(s) \quad \text{for some } 0 < s \leq s_0\}$$

$$= \tilde{P}^{(y_0,s_0)}_{(0,0)}\{\sigma > 0\}$$

$$= \int_{\{\sigma > 0\}}(\tilde{L}_\sigma)^{-1}d\tilde{Q}^{(y_0,s_0)}$$

$$= a^{-1}f(\frac{y_0}{s_0}, \frac{1}{s_0}).$$

□□□

3. From the method of images to the tangent approximation

For the method of images the first exit density of Brownian motion over the boundary $\psi_a(t)$ according to Theorem 1.2 can be expressed as

$$(3.1) \quad p_a(t) = \frac{E(\theta \mid (\psi_a(t), t))}{2t^{3/2}} \; \phi\left(\frac{\psi_a(t)}{\sqrt{t}}\right)$$

with
$$E(\theta \mid (\psi_a(t), t)) = \frac{\int \theta \phi\left(\frac{\theta - \psi_a(t)}{\sqrt{t}}\right) F(d\theta)}{\int \phi\left(\frac{\theta - \psi_a(t)}{\sqrt{t}}\right) F(d\theta)} \;.$$ Here

the boundary function $\psi_a(t)$ satisfies the implicit equation

$$\frac{1}{\sqrt{t}} \phi\left(\frac{\psi_a(t)}{\sqrt{t}}\right) = a^{-1} \int_0^\infty \frac{1}{\sqrt{t}} \phi\left(\frac{\theta - \psi_a(t)}{\sqrt{t}}\right) F(d\theta).$$

Since ψ_a is concave by Lemma 1.1, the subsequent inequality holds for the first exit density

$$(3.2) \quad p_a(t) \leq \frac{\Lambda_a(t)}{t^{3/2}} \; \phi\left(\frac{\psi_a(t)}{\sqrt{t}}\right)$$

with $\Lambda_a(t) = \psi_a(t) - t\psi_a'(t)$ by the following argument. According to the Bachelier-Levy formula, given in the introduction, the right hand side of (3.2) is the first exit density at t of the Brownian motion over the tangent to the curve ψ_a at t. Since the tangent $\psi(t) - (t-u)\psi'(t)$ always lies above the curve $\psi_a(u)$ for $u \leq t$, it is intuitively plausible and easy to prove (see e.g. Lorden (1973) and also the formulas (3.5)-(3.7) below), that the first exit density at the tangent is bigger than that at the curve at time t.

In this section we study the asymptotic behaviour of $p_a(t)$ when $a \to \infty$ for boundaries ψ_a with strong curvature. We shall show that for smooth mixing measures F,

(3.3) $p_a(t) = \dfrac{\Lambda_a(t)}{t^{3/2}} \, \phi(\dfrac{\psi_a(t)}{\sqrt{t}}) \, (1+o(1))$

holds uniformly on intervals $(0,t_1]$ when $a \to \infty$.

This statement is studied from different perspectives, at first from an analytic and further below from a probabilistic point of view.

Compared to statement (3.1) the tangent approximation (3.3) has the advantage that it is easy to calculate. Its quality of approximation for finite situations is discussed in the next section.

We show at first that $\Lambda_a(t)$ can be expressed as a ratio of certain moments. Here we drop the index "a" for a while.

Lemma 3.1:

(3.4) $2\Lambda(t) = \dfrac{E(\theta^2 \mid (\psi(t),t))}{E(\theta \mid (\psi(t),t))} = \dfrac{\int \theta^2 F_{\psi(t),t}(d\theta)}{\int \theta F_{\psi(t),t}(d\theta)}$

where the measure

$$F_{\psi(t),t}(d\theta) = \dfrac{\phi(\dfrac{\theta-\psi_a(t)}{\sqrt{t}}) F(d\theta)}{\int \phi(\dfrac{\theta-\psi_a(t)}{\sqrt{t}}) F(d\theta)}$$

$$= a^{-1} \exp(\theta\psi_a(t)/t - \tfrac{1}{2}\theta^2/t) F(d\theta).$$

Proof: Differentiation of the implicit equation (1.3) for ψ

$$a = \int_0^\infty \exp(\theta\beta(t) - \tfrac{1}{2}\theta^2/t) F(d\theta)$$

with $\beta(t) = \psi(t)/t$ yields

$$0 = \beta'(t) \int_0^\infty \theta \cdot \exp(\theta\beta(t) - \tfrac{1}{2}\theta^2/t) F(d\theta)$$

$$+ \frac{1}{2t^2} \int_0^\infty \theta^2 \exp(\theta\beta(t) - \frac{1}{2}\theta^2/t) F(d\theta)$$

where $\beta'(t) = -\Lambda(t)/t^2$. But rewriting the last equation yields (3.4).

□□□

Combining now the formulas (3.1) and (3.4) yields the representation of the first exit density

$$(3.5) \quad p(t) = \frac{\Lambda(t)}{t^{3/2}} \phi(\frac{\psi(t)}{\sqrt{t}}) \frac{A(t)}{B(t)} \quad \text{with}$$

$$(3.6) \quad \frac{A(t)}{B(t)} = \frac{E(\theta \mid (\psi(t), t))^2}{E(\theta^2 \mid (\psi(t), t))} .$$

By the Cauchy-Schwartz inequality

$$(3.7) \quad 0 < \frac{A(t)}{B(t)} \leq 1 \quad \text{which by (3.5) yields (3.2).}$$

The ratio is exactly one if and only if F consists of a single point mass, which means that ψ_a is a linear boundary.

The subsequent lemma describes the crucial facts which make the following asymptotic considerations possible. Here $T_a = \inf\{t > 0 \mid W(t) \geq \psi_a(t)\}$.

Lemma 3.2: Let $t_1 > 0$ be arbitrary. Then as $a \to \infty$

$$(3.8) \quad P(T_a \leq t_1) \to 0 \quad and$$

$$(3.9) \quad \inf_{0 < t \leq t_1} \frac{\psi_a(t)}{\sqrt{t}} \to \infty .$$

Proof: At first we give that of (3.8). By the definition of ψ_a, $\psi_a(t_1) \to \infty$ as $a \to \infty$.

$$P(T_a \leq t_1) = P(W(t_1) \geq \psi_a(t_1))$$

$$+ \int_{-\infty}^{\psi_a(t_1)} P(T_a < t_1 \mid W(t_1) = x) P(W(t_1) \in dx)$$

$$= 1 - \Phi\left(\frac{\psi_a(t_1)}{\sqrt{t_1}}\right)$$

$$+ \int_{-\infty}^{\psi_a(t_1)} a^{-1} f\left(\frac{x}{t_1}, \frac{1}{t_1}\right) \frac{1}{\sqrt{t_1}} \phi\left(\frac{x}{\sqrt{t_1}}\right) dx$$

by Theorem 1.1. Since $\psi_a(t_1) \to \infty$, the first term converges to zero. Since further for every x, $a^{-1} f\left(\frac{x}{t_1}, \frac{1}{t_1}\right) \to 0$ and since $a^{-1} f\left(\frac{x}{t_1}, \frac{1}{t_1}\right) \leq 1$ for all $x \leq \psi_a(t_1)$, also the second integral converges to zero. This yields (3.8).

To prove (3.9) we note that for $0 < t \leq t_1$

$$1 - \Phi\left(\frac{\psi_a(t)}{\sqrt{t}}\right) = P(W(t) \geq \psi_a(t))$$

$$\leq P(T_a \leq t)$$

$$\leq P(T_a \leq t_1) \to 0 \quad .$$

as $a \to \infty$ by (3.8).

□□□

A similar result holds for the case when F is concentrated on $(-\infty, \infty)$. But then the boundaries can be closed.

To derive the tangent approximation we put the following conditions on the mixing measure: F has a density $f(\theta)$ with the properties

(I) a) there exists a monotone increasing function $h: \mathbb{R}_+ \to [1, \infty)$ and certain positive constants K and v_o

$$\int_{v_o}^{\infty} (1+Kv)^2 h(Kv) \phi(v) dv < \infty$$

such that $f(\theta(1+\alpha)) \leq f(\theta)h(\alpha)$ for all $\alpha \geq 0$.

b) The ratio $f(\theta(1+\epsilon))/f(\theta) \to 1$ uniformly for all θ as $\epsilon \to 0$.

(II) $\quad \sup_{0<y<\infty} F(0,y)/(f(y)e^{y^{2-\delta}}) < \infty$ for $0<\delta<1$.

The conditions (I) and (II) for instance are fulfilled by monotone decreasing densities with exponential tails or by densities which grow like power functions. The following example which has an integrable singularity at $\theta=0$ is also covered by the assumptions (see also Example 4 of Section 2).

$$f(\theta) = \begin{cases} 1/(\theta(\log 1/\theta)(\log_2 1/\theta) \ldots (\log_{n-1} 1/\theta)(\log_n 1/\theta)^{1+\delta}) \\ \qquad\qquad\qquad\qquad\qquad\qquad\qquad\qquad \text{for } 0<\theta \leq e_n \\ f(e_n) \quad \text{for } \theta > e_n \end{cases}$$

Here $\log_2 x = \log(\log x)$, $e_2 = e^e$ etc.

According to Robbins-Siegmund (1970) the corresponding boundary satisfies

$$\psi_a(t) = \Big[2t(\log_2 t^{-1} + 3/2 \log_3 t^{-1} + \sum_{k=4}^{n} \log_k t^{-1} + (1+\delta)\log_{n+1} t^{-1} $$
$$+ \log \tfrac{1}{2}a/\sqrt{\pi} + o(1) \Big]^{1/2} \quad \text{as } t \to 0.$$

Theorem 3.1: Let $t_1 > 0$ be arbitrary. Assume that the prior F satisfies the assumptions (I) and (II). Then

$$(3.10) \quad p_a(t) = \frac{\Lambda_a(t)}{t^{3/2}} \phi(\frac{\psi_a(t)}{\sqrt{t}})(1+o(1))$$

uniformly on $(0,t_1]$ as $a \to \infty$.

Proof: We show that for k=0,1,2

$$(3.11) \quad \int \theta^k \phi\left(\frac{\theta-\psi_a(t)}{\sqrt{t}}\right) F(d\theta) = \sqrt{t}\,(\psi_a(t))^k f(\psi_a(t))(1+o(1))$$

uniformly on $(0,t_1]$ as $a \to \infty$.

Then the equations (3.6) and (3.11) imply that $A(t)/B(t) \to 1$ uniformly on $(0,t_1]$ as $a \to \infty$ which yields the tangent approximation by (3.5). The proof of (3.11) follows by a calculation using Laplace's method. For it we introduce constants $c_a(t) = 2t_1^{(1-\delta)/2}\left(\frac{\sqrt{t}}{\psi_a(t)}\right)^\delta$ with the constant δ of assumption (II). The c_a satisfy $c_a(t) \to 0$ and $c_a(t)\frac{\psi_a(t)}{\sqrt{t}} \to \infty$ uniformly on $(0,t_1]$ as $a \to \infty$ by Lemma 3.2. For k=0,1,2 holds

$$\int_0^\infty \theta^k \phi\left(\frac{\theta-\psi_a(t)}{\sqrt{t}}\right) f(\theta) d\theta =$$

$$= (\psi_a(t))^k \int_0^\infty \left(\frac{\theta}{\psi_a(t)}\right)^k \phi\left(\frac{\theta-\psi_a(t)}{\sqrt{t}}\right) f(\theta) d\theta$$

$$= \psi_a(t)^{k+1} \int_0^\infty (1+u)^k \phi\left(\frac{\psi_a(t)}{\sqrt{t}}u\right) f(\psi_a(t)(1+u)) du$$

where $u = (\theta-\psi_a(t))/\psi_a(t)$.

We split this integral up into three parts

$$= \int_{u<-c_a} + \int_{|u| \le c_a} + \int_{u>c_a} = I + II + III .$$

$$II = \int_{|u| \le c_a} (1+u)^k \phi\left(\frac{\psi_a(t)}{\sqrt{t}}u\right) f(\psi_a(t)(1+u)) du$$

$$= f(\psi_a(t)) \int_{|u| \le c_a} \phi\left(\frac{\psi_a(t)}{\sqrt{t}}u\right) du\,(1+o(1))$$

since $c_a \to 0$ and by assumption (I). Introducing $v = \frac{\psi_a(t)}{\sqrt{t}}u$ yields further

$$(3.12) \quad II = f(\psi_a(t)) \frac{\sqrt{t}}{\psi_a(t)} \int_{|v| \le d_a} \phi(v) dv (1+o(1))$$

$$= f(\psi_a(t)) \frac{\sqrt{t}}{\psi_a(t)} (1+o(1))$$

uniformly on $(0,t_1]$ as $a \to \infty$.

Here $d_a(t) = c_a(t) \dfrac{\psi_a(t)}{\sqrt{t}} \to \infty$ as $a \to \infty$.

The third term can be estimated as follows

$$(3.13) \quad III = \int_{c_a}^{\infty} (1+u)^k \phi(\frac{\psi_a(t)}{\sqrt{t}} u) f(\psi_a(t)(1+u)) du$$

$$\le f(\psi_a(t)) \int_{c_a}^{\infty} (1+u)^k h(u) \phi(\frac{\psi_a(t)}{\sqrt{t}} u) du$$

$$= f(\psi_a(t)) \sqrt{t} (\psi_a(t))^{-1} \int_{d_a}^{\infty} (1+ \frac{\sqrt{t}}{\psi_a(t)} v)^k h(\frac{\sqrt{t}}{\psi_a(t)} v) \phi(v) dv$$

$$\le f(\psi_a(t)) \sqrt{t} (\psi_a(t))^{-1} \int_{d_a}^{\infty} (1+Kv)^k h(Kv) \phi(v) dv .$$

By assumption (I) it follows that

$$III = f(\psi_a(t)) \sqrt{t} (\psi_a(t))^{-1} \cdot o(1)$$

uniformly on $(0,t_1]$ as $a \to \infty$.

$$(3.14) \quad I \le \int_{0}^{(1-c_a)\psi_a(t)} \phi(\frac{\theta - \psi_a(t)}{\sqrt{t}}) f(\theta) d\theta$$

$$\le \phi(- \frac{c_a \psi_a(t)}{\sqrt{t}}) F(0, \psi_a(t))$$

$$= \frac{\psi_a(t)}{\sqrt{t}} \; \phi(\frac{c_a\psi_a(t)}{\sqrt{2t}}) \phi(-\frac{c_a\psi_a(t)}{\sqrt{2t}}) \frac{F(0,\psi_a(t))}{f(\psi_a(t))} \; \frac{\sqrt{t}}{\psi_a(t)} \; f(\psi_a(t))$$

$$= o(1) \; e^{-\psi_a(t)^{2-\delta}} \; \frac{F(0,\psi_a(t))}{f(\psi_a(t))} \; \frac{\sqrt{t}}{\psi_a(t)} \; f(\psi_a(t))$$

$$= o(1) \; \frac{\sqrt{t}}{\psi_a(t)} \; f(\psi_a(t))$$

by assumption (II).

(3.12), (3.13) and (3.14) together yield (3.11) and thus the theorem.

□□□

Combining the statements (3.1) and (3.11) yields the following result.

Corollary 3.1:

$$p_a(t) = \frac{\psi_a(t)}{2t^{3/2}} \; \phi(\frac{\psi_a(t)}{\sqrt{t}}) \; (1+o(1)).$$

Comparing this statement with (3.10) shows that $\Lambda_a(t) = \frac{1}{2} \psi_a(t)(1+o(1))$, which means that $\psi_a(t)$ is nearly parabolic. Thus boundaries which grow like t^α, $\alpha < \frac{1}{2}$ are ruled out by the assumptions. In fact the measures F of Example 5 of Section 2, which yield such boundaries, do not satisfy assumption (I). Nevertheless those boundaries will be covered by the following considerations.

The preceding analytic derivation does not offer much insight why the tangent approximation (3.10) holds. Therefore we give in the sequel an intuitive heuristic argument which shows that statement (3.10) is quite natural from the probabilistic point of view. Further below we shall make this argument rigorous.

The heuristic argument is this: for the tangent-approximation (3.10) to hold it is necessary that the mass of those paths of Brownian motion which cross the curve ψ_a much earlier than t but the tangent at first at t, is small compared to that of paths which cross the curve at time t for the first time.

To make this statement more precise let us consider instead of Brownian motion the Brownian Bridge with endpoints $(0,0)$ and $(\psi_a(t),t)$. If for every $\varepsilon > 0$ small

$$(3.15) \quad P(T_a \leq (1-\varepsilon)t \mid W(t)=\psi_a(t)) = o(1)$$

as $a \to \infty$, it is intuitively clear and we shall show this at the end of this section that the tangent approximation will hold. But why holds (3.15)? To answer this question let $\tilde{W}_0$ denote the Brownian Bridge with endpoints $(0,0)$ and $(\psi_a(t),t)$. Then

$$(3.16) \quad \sup_{0 < v \leq 1} \left| \frac{\tilde{W}_0(vt)}{\psi_a(t)} - v \right| \to 0$$

uniformly on $(0,t_1]$ as $a \to \infty$.

This can be seen as follows. Let $W_0(vt) = \tilde{W}_0(vt) - v\psi_a(t)$. Then by the scaling property of the Brownian Bridge

$$\sup_{0 < v \leq 1} |\tilde{W}_0(vt)/\psi_a(t) - v| = \sup_{0 < v \leq 1} \left| \frac{W_0(vt)/\sqrt{t}}{\psi_a(t)/\sqrt{t}} \right|$$

$$\overset{L}{=} \sup_{0 < v \leq 1} \left| \frac{W_0(v)}{\psi_a(t)/\sqrt{t}} \right| \quad .$$

But this expression converges to zero uniformly on $(0,t_1]$ as $a \to \infty$ by (3.9).

Thus statement (3.16) describes the fact that if the endpoint is high, the Brownian Bridge takes nearly the shortest way between $(0,0)$ and $(\psi_a(t),t)$, which is along the ray $\frac{u}{t} \psi_a(t)$. If the boundary $\psi_a(u)$ for $u < t$ is high relative to the ray $\frac{u}{t} \psi_a(t)$, then one might expect that

(3.15) holds.* Here height is measured in units of the standard deviation of the Brownian Bridge W_o. In fact under appropriate conditions on
F

$$(3.17) \quad \inf_{0<u\le t(1-\varepsilon)} \sqrt{\frac{t}{u(t-u)}} \left(\psi_a(u) - \frac{u}{t}\psi_a(t)\right) \to \infty \quad \text{holds} \ .$$

We show this below. But (3.17) does not imply (3.15) because of the wild
fluctuations of the Brownian Bridge near t=0.

The following considerations make the preceding remarks precise and take
into account the behaviour of the Brownian Bridge near zero. For this we
put the following assumption on the prior F:

(III) It exists a $\kappa>0$ and a constant M>0 with

$$(3.18) \quad \varliminf_{a\to\infty} \inf_{0<t\le t_1} F_{\psi_a(t),t}(M\psi_a(t),\infty) > \kappa$$

$$\text{where} \quad F_{\psi_a(t),t}(d\theta) = \frac{\phi\left(\dfrac{\theta-\psi_a(t)}{\sqrt{t}}\right)F(d\theta)}{\int\phi\left(\dfrac{\theta-\psi_a(t)}{\sqrt{t}}\right)F(d\theta)} \ .$$

Assumption (III) for instance is fulfilled if F has a monotone decreasing
density and assumption (II) holds.

The following consequence of assumption (III) is crucial for the subsequent considerations: there exists a constant L>0 with

$$(3.19) \quad L\psi_a(t) \le \Lambda_a(t) \quad \text{for all } 0<t\le t_1 \ .$$

---

*These considerations are obviously closely related to the large devi-
ation theory as it is described in Borovkov (1967) and Varadhan (1966):
for quick pathes the functional $\int_0^t (\dot{x}(u))^2 du$ has to be small. This
functional also characterizes the set of limit points of the functional
law of the iterated logarithm due to Strassen (1964). Varadhan (1973)
showed that Strassen's result can be obtained from the general large
deviation theory for quick paths.

This inequality follows from the next inequality by using (3.1) and (3.2):

$$M\psi_a(t)\kappa \le \int_{M\psi_a(t)}^{\infty} \theta F_{\psi_a(t),t}(d\theta) \le \Lambda_a(t) .$$

We note that (3.19) is not satisfied if F has compact support on $(0,\infty)$. Then $\Lambda_a(t)$ stays bounded but $\psi_a(t)$ tends to infinity.

Let $0<t_1$ be arbitrary. For $t\epsilon(0,t_1]$ let s be defined by the solution of the implicit equation $s=t(1-(\frac{s}{\psi_a(s)^2})^\delta)$ with $0<\delta<\frac{1}{2}$. Since by Lemma 3.2 $\inf_{0<u\le t_1} \psi_a(u)^2/u\to\infty$, it holds $s/t \to 1$ for $a \to \infty$ uniformly on $(0,t_1]$.

Theorem 3.2: Let the mixing measure F satisfy assumption (III). *Then*

(3.20) $P(T_a\le s|W(t)=\psi_a(t))=o(1)$

uniformly on $(0,t_1]$ *as* $a \to \infty$.

Since on the other hand $P(T_a<t|W(t)=\psi_a(t))=1$ by the law of the iterated logarithm, the theorem states that almost all crossings happen just before t.

Proof: We drop the index a. Theorem 1.1 yields

(3.21) $P(T\le s|W(t)=\psi(t))=P(W(s)\ge\psi(s)|W(t)=\psi(t))$

$$+ \int_{-\infty}^{\psi(s)} P(T\le s|W(s)=x)P(W(s)\epsilon dx|W(t)=\psi(t))$$

$$=1-\Phi\left(\sqrt{\frac{t}{s(t-s)}}\ (\psi(s)-\frac{s}{t}\psi(t))\right)$$

$$+ \int_{-\infty}^{\psi(s)} a^{-1}f(\frac{x}{s},\frac{1}{s})\sqrt{\frac{t}{s(t-s)}}\ \phi\left(\sqrt{\frac{t}{s(t-s)}}(x-\frac{s}{t}\psi(t))\right)dx .$$

Here we also use that the distribution of the underlying Brownian
Bridge is given by

(3.22) $\quad P(W(u) \in dx \mid W(t) = \psi(t)) = dx \dfrac{\frac{1}{\sqrt{u}}\phi(\frac{x}{\sqrt{u}})\frac{1}{\sqrt{t-u}}\phi(\frac{\psi(t)-x}{\sqrt{t-u}})}{\frac{1}{\sqrt{t}}\phi(\frac{\psi(t)}{\sqrt{t}})}$

$$= dx \sqrt{\frac{t}{u(t-u)}}\ \phi\left(\sqrt{\frac{t}{u(t-u)}}\ (x-\tfrac{u}{t}\psi(t))\right).$$

Now we estimate the first term of (3.21). By concavity

(3.23) $\quad \dfrac{\psi(u)-\frac{u}{t}\psi(t)}{(t-u)} = \dfrac{\Lambda(u)}{t} + \dfrac{u}{t}\left[\psi'(u)-\dfrac{\psi(t)-\psi(u)}{(t-u)}\right] \geq \dfrac{\Lambda(u)}{t}$

for u<t. This inequality and statement (3.19) yield the estimate

(3.24) $\quad \sqrt{\dfrac{t}{s(t-s)}}\ (\psi(s)-\tfrac{s}{t}\psi(t)) \geq \sqrt{\dfrac{t}{s}}\ \dfrac{\Lambda(s)}{t}\ \sqrt{t-s}$

$$\geq L\sqrt{\dfrac{t}{s}}\ \dfrac{\psi(s)}{t}\ \sqrt{t(\dfrac{s}{\psi(s)^2})}^{\delta}$$

$$= L(\dfrac{\psi(s)}{\sqrt{s}})^{1-\delta}\ .$$

The right hand side converges to infinity uniformly on $(0,t_1]$ as $a \to \infty$
by (3.9). Then (3.24) yields that the first term of (3.21) converges to
zero. The second term of (3.21) we split up into two parts
$\int_{-\infty}^{x_s} + \int_{x_s}^{\psi(s)}$ where $x_s < \psi(s)$. We choose $x_s = \tfrac{s}{t}\psi(t) + K_s\sqrt{\dfrac{s(t-s)}{t}}$ with
$K_s = (\dfrac{\psi(s)}{\sqrt{s}})^{\delta}$. Then (3.24) yields

(3.25) $\quad \sqrt{\dfrac{t}{s(t-s)}}(\psi(s)-x_s) = \sqrt{\dfrac{t}{s(t-s)}}(\psi(s)-\tfrac{s}{t}\psi(t))-K_s$

$$\geq L(\dfrac{\psi(s)}{\sqrt{s}})^{1-\delta}(1-L^{-1}(\dfrac{\sqrt{s}}{\psi(s)})^{1-2\delta})$$

$$\geq \frac{1}{2} L \left(\frac{\psi(s)}{\sqrt{s}}\right)^{1-\delta}$$

uniformly on $(0,t_1]$ as $a \to \infty$. Since for $x < \psi(s)$ $\quad a^{-1}f(\frac{x}{s},\frac{1}{s}) \leq 1$, the integral

$$\int_{x_s}^{\psi(s)} a^{-1}f(\frac{x}{s},\frac{1}{s})\sqrt{\frac{t}{s(t-s)}}\, \phi\left(\sqrt{\frac{t}{s(t-s)}}(x-\frac{s}{t}\psi(t))\right)dx$$

can now be estimated by $1-\phi\left(\sqrt{\frac{t}{s(t-s)}}(x_s-\frac{s}{t}\psi(t))\right)$ which converges to zero by the choice of x_s. It is left to estimate

$$\int_{-\infty}^{x_s} a^{-1}f(\frac{x}{s},\frac{1}{s})\sqrt{\frac{t}{s(t-s)}}\, \phi\left(\sqrt{\frac{t}{s(t-s)}}(x-\frac{s}{t}\psi(t))\right)dx.$$

Since $f(\frac{x}{s},\frac{1}{s}) \leq f(\frac{x_s}{s},\frac{1}{s})$ for $x \leq x_s$, it is sufficient to estimate $f(\frac{x_s}{s},\frac{1}{s})$.

$$a^{-1}f(\frac{x_s}{s},\frac{1}{s}) = a^{-1}\int \exp(\theta x_s/s - \frac{1}{2}\theta^2/s)F(d\theta)$$

$$= a^{-1}\int \exp\left[-\frac{\theta}{2s}(\psi(s)-x_s)\right]\exp(\theta\psi(s)/s-\frac{1}{2}\theta^2/s)F(d\theta)$$

$$= a^{-1}\left(\int_{\theta \leq \theta_s} + \int_{\theta > \theta_s}\right)$$

where $\theta_s = \left(\frac{\sqrt{s}}{\psi(s)}\right)^\gamma \sqrt{s}$ with $0 < \gamma < 1-2\delta$. The integral

$$a^{-1}\int_{\theta > \theta_s} \leq \exp\left(-\frac{L}{2}\left(\frac{\psi(s)}{\sqrt{s}}\right)^{1-2\delta-\gamma}\right)$$

since by (3.25) $s^{-1}(\psi(s)-x_s) \geq \frac{1}{2}L\left(\frac{\psi(s)}{\sqrt{s}}\right)^{1-2\delta} s^{-\frac{1}{2}}$ and since

$$a^{-1}\int \exp(\theta\psi(s)/s-\frac{1}{2}\theta^2/s)F(d\theta) = 1 .$$

Therefore it converges to zero by (3.9). To estimate the other part

$\int\limits_{\theta<\theta_s}$ it is sufficient to show that

$$a^{-1} \int\limits_{0}^{\theta_s} \exp(\theta\psi(s)/s - \tfrac{1}{2}\theta^2/s) F(d\theta) = o(1).$$

Since $\sup\limits_{0<s\leq t_1} F(0,\theta_s) \to 0$ by the definition of θ_s and since

$$\frac{\int\limits_{0}^{\theta_s} \exp(\theta\psi(s)/s) F(d\theta)}{\int\limits_{\theta_s}^{\infty} \exp(\theta\psi(s)/s)) \exp(-\tfrac{1}{2}\theta^2/s) F(d\theta)} \leq \frac{F(0,\theta_s)}{\int\limits_{\theta_s}^{\infty} \exp(-\tfrac{1}{2}\theta^2/s) F(d\theta)}$$

it follows from the equation $a = \int\limits_{0}^{\infty} \exp(\theta\psi(s)/s - \tfrac{1}{2}\theta^2/s) F(d\theta)$ that

$$a = \int\limits_{\theta_s}^{\infty} \exp(\theta\psi(s)/s - \tfrac{1}{2}\theta^2/s) F(d\theta)(1+o(1)).$$

This implies the desired estimate and completes the proof.

□□□

In the sequel we study the derivation of the tangent approximation from Theorem 3.2. To explain this we need several integral equations, which we also will use in the next section and which we introduce first.

Let ψ be a smooth boundary such that the density p of the first exit distribution of Brownian motion over ψ exists. (According to Strassen (1967) a continuous differentiable boundary has a continuous first exit density if it is an upper class function at zero.) Let T denote the corresponding first exit time. Then by the law of the iterated logarithm it follows that

(3.26) $P(T<t \mid W(t) = \psi(t)) = 1$.

We can write this equation more explicitly by using the first exit density of the Brownian Bridge, which by formula (3.22) and the strong

Markov property of the Brownian motion is given by

$$p(u) \frac{1}{\sqrt{t-u}} \phi(\frac{\psi(t)-\psi(u)}{\sqrt{t-u}}) \left[\frac{1}{\sqrt{t}} \phi(\frac{\psi(t)}{\sqrt{t}})\right]^{-1} \ .$$

Thus (3.26) can be expressed as

$$1 = \int_0^t p(u) \frac{\frac{1}{\sqrt{t-u}} \phi(\frac{\psi(t)-\psi(u)}{\sqrt{t-u}})}{\frac{1}{\sqrt{t}} \phi(\frac{\psi(t)}{\sqrt{t}})} \, du \ ,$$

which is equivalent to the following integral equation (c.f. Fortet (1943) and Durbin (1971)):

$$(3.27) \qquad \frac{1}{\sqrt{t}} \phi(\frac{\psi(t)}{\sqrt{t}}) = \int_0^t p(u) \frac{1}{\sqrt{t-u}} \phi(\frac{\psi(t)-\psi(u)}{\sqrt{t-u}}) \, du \ .$$

The next integral equation, which we derive, is due to Durbin (1985) and Ferebee (1982). For it we assume additionally that ψ is concave. Let $\Lambda(t) = \psi(t) - t\psi'(t)$.

$$(3.28) \qquad p(t) = \frac{\Lambda(t)}{t^{3/2}} \phi(\frac{\psi(t)}{\sqrt{t}}) - \int_0^t p(u) \frac{\psi(t)-\psi(u)-(t-u)\psi'(t)}{(t-u)^{3/2}} \phi(\frac{\psi(t)-\psi(u)}{\sqrt{t-u}}) \, du$$

This equation has an intuitive geometrical meaning for concave boundaries. Let

$$S = \inf\{u > 0 \mid W(u) \geq \psi(t) - (t-u)\psi'(t)\}$$

denote the first exit time of Brownian motion over the tangent to the curve ψ at t. By concavity holds $\psi(t) - (t-u)\psi'(t) \geq \psi(u)$ for all $u \leq t$ and therefore for all paths with $T \leq t$, holds $T \leq S$. The Bachelier-Levy formula yields

$$(3.29) \qquad P(S \in dt) = \frac{\Lambda(t)}{t^{3/2}} \phi(\frac{\psi(t)}{\sqrt{t}}) \, dt$$

and together with the strong Markov property

$$(3.30) \quad P(T<S, \ S\epsilon dt) = \int_0^t P(T\epsilon du, \ S\epsilon dt)$$

$$= \int_0^t du \ p(u) \ \frac{\psi(t)-\psi(u)-(t-u)\psi'(t)}{(t-u)^{3/2}} \ \phi(\frac{\psi(t)-\psi(u)}{\sqrt{t-u}})dt.$$

Substituting the equations (3.29) and (3.30) into (3.28) shows that equation (3.28) can be expressed as

$$(3.31) \quad P(T\epsilon dt) = P(S\epsilon dt)-P(S\epsilon dt,T<S) \quad ,$$

which is obviously true.

By combining the equations (3.27) and (3.28) we can derive further integral equations for concave functions, for instance the one with which we shall show the tangent approximation:

$$(3.32) \quad p(t) = \int_0^t p(u) \ \frac{\psi(u)-\frac{u}{t}\psi(t)}{(t-u)^{3/2}} \ \phi(\frac{\psi(t)-\psi(u)}{\sqrt{t-u}})du \quad .$$

This equation can be derived as follows. It is obvious that (3.28) can be written as

$$(3.33) \quad p(t) = \frac{\Lambda(t)}{t^{3/2}} \ \phi(\frac{\psi(t)}{\sqrt{t}}) - \int_0^t \frac{\psi(t)-\psi(u)}{(t-u)^{3/2}} \ \phi(\frac{\psi(t)-\psi(u)}{\sqrt{t-u}})p(u)du$$

$$+ \ \psi'(t)\int_0^t \frac{1}{(t-u)^{1/2}} \ \phi(\frac{\psi(t)-\psi(u)}{\sqrt{t-u}})p(u)du \quad .$$

An application of the equation (3.27) on the last term of (3.33) yields

$$(3.34) \quad p(t) = \frac{\psi(t)}{t^{3/2}} \ \phi(\frac{\psi(t)}{\sqrt{t}}) - \int_0^t \frac{\psi(t)-\psi(u)}{(t-u)^{3/2}} \ \phi(\frac{\psi(t)-\psi(u)}{\sqrt{t-u}})p(u)du.$$

A further application of the equation (3.27) on the first term of (3.34) leads to (3.32).

The equation (3.34) is interesting by itself. For a monotone increasing function ψ it has an intuitive probabilistic meaning, which can be expressed by

(3.35) $\quad P(T \in dt) = P(U \in dt) - P(U \in dt, T < U)$

where $\quad U = \inf\{u > 0 \mid W(u) \geq \psi(t)\}$.

For monotone increasing functions one derives (3.28) and (3.32) by combining the equations (3.27) and (3.34).

We note that the probabilistic meaning of (3.34) for concave decreasing functions is unclear; the same can be said for the equation (3.28) with convex increasing functions. Nevertheless Ferebee (1982) has shown with analytical methods that (3.28) holds without any order restrictions on ψ, which implies the same for the equations (3.32) and (3.34). But his arguments offer no simple probabilistic interpretation of the equations.

We now come back to the derivation of the tangent approximation. Because of the inequality

(3.2) $\quad p_a(t) \leq \dfrac{\Lambda_a(t)}{t^{3/2}} \phi\left(\dfrac{\psi_a(t)}{\sqrt{t}}\right)$,

we have only to show the converse inequality as $a \to \infty$. In the sequel we omit the subindex a. By Theorem 3.2 we have for s given by the solution of the implicit equation $\quad s = t\left(1 - \left(\dfrac{s}{\psi(s)^2}\right)^{\delta}\right)$ with $0 < \delta < \dfrac{1}{2}$

(3.36) $\quad 1 + o(1) \leq P(s < T < t \mid W(t) = \psi(t))$

$$= \int_s^t p(u) \; \frac{\frac{1}{\sqrt{t-u}} \phi\left(\frac{\psi(t)-\psi(u)}{\sqrt{t-u}}\right)}{\frac{1}{\sqrt{t}} \phi\left(\frac{\psi(t)}{\sqrt{t}}\right)} \; du \; .$$

Since by (3.23) $\dfrac{\psi(u)-\frac{u}{t}\psi(t)}{t-u} \geq \dfrac{\Lambda(u)}{t} \geq \dfrac{\Lambda(s)}{t}$ we get from (3.36)

(3.37) $\dfrac{\Lambda(s)}{t^{3/2}} \phi(\dfrac{\psi(t)}{\sqrt{t}})(1+o(1)) \leq \displaystyle\int_s^t p(u) \dfrac{\psi(u)-\frac{u}{t}\psi(t)}{(t-u)^{3/2}} \phi(\dfrac{\psi(t)-\psi(u)}{\sqrt{t-u}}) du$

$$\leq \int_o^t p(u) \dfrac{\psi(u)-\frac{u}{t}\psi(t)}{(t-u)^{3/2}} \phi(\dfrac{\psi(t)-\psi(u)}{\sqrt{t-u}}) du$$

$$= p(t) \ .$$

Here we use the concavity of ψ, which implies $\psi(u)-\frac{u}{t}\psi(t)>0$ for $0<u<t$ and equation (3.32). This finally yields

(3.38) $\dfrac{\Lambda_a(s)}{t^{3/2}} \phi(\dfrac{\psi_a(t)}{\sqrt{t}})(1+o(1)) \leq p_a(t)$

uniformly on $(0,t_1]$ as $a \to \infty$.

Under a condition similar to (I) and under (III)

(3.39) $|\Lambda_a(s)/\Lambda_a(t) - 1| \to 0$

uniformly on $(0,t_1]$ as $a \to \infty$. Then (3.38) together with (3.2) yield the tangent approximation. We omit the details.

In the next section we study the tangent approximation in general using several of the ideas presented in this section. There we shall put assumptions on ψ which are quite similar to the ones which turned out here as necessary conditions for the method of images. Unfortunately the results of the next section are somewhat too weak to cover the tangent approximation for the method of images in general. We shall come back to this point in a remark which follows Theorem 4.1.

4. The tangent approximation

In this section we study the tangent approximation in general and derive it for a broad class of boundaries. It will be shown that the tangent approximation holds uniformly over intervals if the boundaries recede to infinity. Therefore by integrating out the densities, approximations for the first exit probabilities can be derived. The sets on which the tangent approximation holds can be finite intervals or the whole real line. This depends on the fact, whether the boundaries belong to the upper or lower class at infinity. All the results about the tangent approximation hold uniformly over all drift directions. A refinement of the tangent approximation, a second order approximation due to Jennen is also given and the quality of the approximation is discussed for some examples.

Let $\{\psi_a ; a \in \mathbb{R}_+\}$ denote a set of positive, monotone increasing, continuously differentiable functions on $\mathbb{R}_+ = (0, \infty)$. (The assumption of monotonicity is just for simplicity, it can be removed). The first exit time of standard Brownian motion $W(t)$ over ψ_a is defined by

$$(4.1) \qquad T_a = \inf\{t > 0 \mid W(t) \geq \psi_a(t)\} .$$

According to Strassen (1967) the distribution of the stopping time T_a has a continuous density p_a on $(0, \infty)$ for each a provided that ψ_a is an upper-class function at zero. Let $\Lambda_a(t) = \psi_a(t) - t\psi_a'(t)$ denote the intercept of the tangent at t to the curve ψ_a on the space-axis. The following theorem is very similar to Theorem 1 of Jennen-Lerche (1981). A proof of it is given at the end of this section.

Theorem 4.1: Let $0 < t_1 \leq \infty$ and $0 < \alpha < 1$. *Assume that*

(I) $\qquad P(T_a < t_1) \to 0 \quad as \quad a \to \infty,$

(II) $\qquad \psi_a(t)/t^\alpha$ *is monotone decreasing in t for each a,*

(III) *for every* $\varepsilon > 0$ *there exists a* $\delta > 0$ *such that for all a*

$$|\psi_a'(s)/\psi_a'(t) - 1| < \varepsilon \quad if \quad |s/t - 1| < \delta$$

for $s,t \in (0,t_1)$.

Then

$$(4.2) \qquad P_a(t) = \frac{\Lambda_a(t)}{t^{3/2}} \phi(\frac{\psi_a(t)}{\sqrt{t}})(1+o(1))$$

uniformly on $(0,t_1)$ *as* $a \to \infty$.

Integration yields the following corollary.

Corollary 4.1:

$$P(T_a < t) = \int_0^t \frac{\Lambda_a(u)}{u^{3/2}} \phi(\frac{\psi_a(u)}{\sqrt{u}}) du (1+o(1))$$

uniformly on $(0,t_1)$ *as* $a \to \infty$.

We add several remarks to the theorem. At first we note that the tangent approximation is a purely local approximation. The quantity $\frac{\Lambda_a(t)}{t^{3/2}} \phi(\frac{\psi_a(t)}{\sqrt{t}})$ is usually not a probability density (except for straight lines).

The assumptions of the theorem are a little bit too strong to cover the method of images in general as described in the last section. Essentially assumption (II) is too strong.

Assumption (I) can easily be checked by using the inequality

$$P(T_a < t_1) \leq \int_0^{t_1} \frac{\psi_a(t)}{t^{3/2}} \phi(\frac{\psi_a(t)}{\sqrt{t}}) dt$$

which holds for monotone functions ψ_a.

The case $t_1 = \infty$ is included. Example (i) (below) is of this type. The other examples satisfy the conditions of Theorem 4.1 on finite intervals:

(4.3)   (i) $\psi_a(t) = \sqrt{(t+1)(2a+\log(t+1))}$,

(ii) $\psi_a(t) = at^\alpha$, $\alpha < \frac{1}{2}$,

(iii) $\psi_a(t) = \sqrt{2(r+at)}$, $r > 0$,

(iv) $\psi_a(t) = \sqrt{a}\psi(t/a)$, ψ a fixed function.

The last example shows that Strassen's result on the tangent approximation (Theorem 3.5 of Strassen (1967)) is contained as a special case of
Theorem 4.1.

The question arises: for which functions is statement (4.2) true on
$\mathbb{R}_+$? In answering this one has to distinguish between upper and lower
class functions at infinity (ψ is an upper or lower class function of
the standard Brownian motion $W(t)$ at infinity according as $P(W(t) < \psi(t)$
for t large)=1 or 0). For upper class functions the answer to our question is affirmative, for lower class functions it is negative. The case
of upper class functions is covered by Theorem 4.1. A typical example is
given by (4.3)(i). To illustrate what happens in the case of lower class
functions let us consider the boundary of example (iii), $\psi_a(t) = \sqrt{2(r+at)}$.
For it by Corollary 4.1 holds

(4.4) $P(T < t_1) = \sqrt{a}e^{-a} \frac{1}{2\sqrt{\pi}} \int_0^{t_1} e^{-r/t} \frac{dt}{t}(1+o(1))$.

The integral on the right hand side tends to infinity as $t_1 \to \infty$. Thus
the tangent approximation cannot be true on the whole axis. In fact for
this boundary the assumptions (II) and (III) of Theorem 4.1 hold on $\mathbb{R}_+$
but not (I) since $P(T_a < \infty) = 1$ for all $a > 0$.

Nevertheless the intervals on which (4.2) holds uniformly, can grow with
a. Therefore it is natural to ask: for which functions $a \rightsquigarrow h_a$ with
$\lim_{a \to \infty} h_a = \infty$, does statement (4.2) hold on $I_a = (0, h_a)$?

To answer this question, we rescale the problem to the interval (0,1)
by using the space-time transformation

(4.5) $s = t/h_a$, $y = x/\sqrt{h_a}$.

We also introduce the new assumption

(I') $P(T_a < h_a) \to 0$ as $a \to \infty$.

If the assumptions (I') and (II) and (III) hold on $\mathbb{R}_+$ for the original functions $\psi_a(t)$, then they are fulfilled for their transforms $\eta_a(s) = \psi_a(h_a s)/\sqrt{h_a}$ on $(0,1)$. Also the quotient of the density and its approximation remains invariant under the transformation. That is, if $f_a(s)$ denotes the density of the first exit distribution over η_a, then

$$p_a(t) \left[\frac{\psi_a(t) - t\psi_a'(t)}{t^{3/2}} \, \phi\left(\frac{\psi_a(t)}{\sqrt{t}}\right) \right]^{-1}$$

$$= f_a(t) \left[\frac{\eta_a(t) - t\eta_a'(t)}{t^{3/2}} \, \phi\left(\frac{\eta_a(t)}{\sqrt{t}}\right) \right]^{-1} \quad ,$$

since $f_a(s) = h_a p_a(t)$. Therefore the tangent approximation (4.2), which holds for η_a by Theorem 4.1, carries over to the tangent approximation for ψ_a on $(0, h_a)$. This yields the following result.

Theorem 4.2: Let h_a be a function with $\lim\limits_{a \to \infty} h_a = \infty$ and let the assumptions (II) and (III) be fulfilled on $\mathbb{R}_+$. Furthermore assume

(I') $P(T_a < h_a) \to 0$ as $a \to \infty$.

Then $p_a(t) = \dfrac{\Lambda_a(t)}{t^{3/2}} \, \phi\left(\dfrac{\psi_a(t)}{\sqrt{t}}\right)(1 + o(1))$

uniformly on $I_a = (0, h_a)$ as $a \to \infty$.

The converse of Theorem 4.2 is also true: in the next section under further assumptions we shall show that (I') is necessary for the tangent approximation to hold on I_a (see Theorem 5.5).

Example 1: $\psi_a(t) = \sqrt{2(r + at)}$ for $t > 0$.
For the first exit density holds

$$p_a(t) = \frac{\sqrt{a}}{2\sqrt{\pi t}} \, e^{-(a+r/t)} (1+o(1))$$

according to Theorem 4.1. From Corollary 4.1 we get

$$(4.6) \quad P(T_a < t_1) = \int_0^{t_1} \frac{\Lambda_a(t)}{t^{3/2}} \, \phi\left(\frac{\psi_a(t)}{\sqrt{t}}\right) dt \, (1+o(1))$$

$$= \sqrt{a} \, e^{-a} \, \frac{1}{2\sqrt{\pi}} \int_0^{t_1} e^{-r/t} \frac{dt}{t} \, (1+o(1))$$

$$= \sqrt{a} \, e^{-a} \, \frac{1}{\sqrt{\pi}} \int_{\sqrt{2/t_1}}^{\infty} e^{-\frac{rx^2}{2}} \frac{dx}{x} \, (1+o(1)).$$

This is the one-sided version of Siegmund's result discussed in Example 6 of Section 2. According to Theorem 4.2 statement (4.6) remains true if we replace the constant t_1 by a function h_a for which the right hand side of (4.6) tends to zero. This follows since the functions ψ_a are concave and therefore the tangent approximation dominates the first exit density which implies (I'). For h_a we can take $h_a = \exp(e^a a^{-\alpha})$ with $\alpha > \frac{1}{2}$.

Example 2: $\psi_{a\pm}(t) = \pm\sqrt{t(\log\frac{a^2}{t})}$ for $0 < t \le t_1$ (two-sided boundary). This is one of the few examples for which the exact density is known (cf. Example 4 of Section 1). For the upper branch ψ_+ it is equal to

$$p_+(t) = \frac{\psi_+(t)}{2t^{3/2}} \, \phi\left(\frac{\psi_+(t)}{\sqrt{t}}\right) \qquad \text{(here we omit the index a).}$$

The approximation (4.2) is given by

$$(4.7) \quad p_+(t) = \left(\frac{\psi_+(t)}{2t^{3/2}} + \frac{1}{2t^{1/2}\psi_+(t)}\right)\phi\left(\frac{\psi_+(t)}{\sqrt{t}}\right)(1+o(1)).$$

It obviously overestimates the true density. For a better approximation see Example 2s following Theorem 4.4. The quality of these approximations is discussed at the end of the section.

Example 3: $\psi_a(t) = at^\alpha$ for $t \geq 0$ with $\alpha > \frac{1}{2}$.
The first exit density is equal to

$$p_a(t) = \frac{1-\alpha}{\sqrt{2\pi}} \, at^{\alpha-\frac{3}{2}} \exp(-a^2/2t^{1-2\alpha})(1+o(1)).$$

For the distribution function holds

$$P(T_a < t_1) = \frac{2(1-\alpha)}{1-2\alpha}\left[1 - \Phi(at_1^{\alpha-1/2})\right](1+o(1)).$$

Theorem 4.2 holds with $h_a = a^\beta$, $\beta < 2/(1-2\alpha)$. For $\alpha > 1/2$ one cannot stop immediately at zero. Nevertheless similar results are true (cf. Jennen-Lerche (1981)).

Example 4: $\psi_a(t) = ((t+1)(2a+\log(t+1)))^{1/2}$ for $t \geq 0$.
(See also Example 3 of Section 2.)

$$P(T < h_a t_1) = e^{-a}(1 - \Phi((2a+\log(h_a t_1 + 1))/h_a t_1)^{1/2}))(1+o(1)).$$

For $h_a = a$ we have

$$P(T_a < at_1) = e^{-a}(1 - \Phi(\sqrt{2/t_1}))(1+o(1))$$

which coincides with equation (35) of Lai-Siegmund (1977). Furthermore (4.7) yields $P(T_a < \infty) = \frac{1}{2}e^{-a}(1+o(1))$.

A simple but important relation makes it possible to extend the previous results to the case of Brownian motion with drift. The approximation (4.2) of Theorem 4.1 states that the ratio of the density to its approximation tends to one. A multiplication of the density and its approximation by the same factor does not change the ratio. To obtain the first exit density and its approximation in the case of Brownian motion with drift, one has to multiply the corresponding quantities of drift-less Brownian motion by the Radon-Nikodym derivative. Let $p_{\theta,a}$ denote the density of the distribution of T_a where

$$T_a = \inf\{t>0 \,|\, W(t) \geq \psi_a(t)\}$$

with $W(t)$ the Brownian motion with drift θ. Then by (1.14)

$$p_{\theta,a}(t) = \exp(\theta\psi_a(t) - \tfrac{1}{2}\theta^2 t)\, p_a(t).$$

We get for the ratio of the density to the approximation the following expression, where the right hand side does not depend on θ.

$$p_{\theta,a}(t)\left[\frac{\Lambda_a(t)}{t^{3/2}}\exp(-(\psi_a(t)-\theta t)^2/2)\right]^{-1}$$

$$= p_a(t)\left[\frac{\Lambda_a(t)}{t^{3/2}}\exp(-\psi_a(t)^2/2t)\right]^{-1}.$$

This leads to the following result.

Theorem 4.3: *Let the conditions* (I'), (II) *and* (III) *on* $\mathbb{R}_+$ *hold. Then*

$$p_{\theta,a}(t) = \frac{\Lambda_a(t)}{t^{3/2}}\,\phi\left(\frac{\psi_a(t)-\theta t}{\sqrt{t}}\right)(1+o(1))$$

uniformly on $(0, h_a)$ *and uniformly for all* $\theta \in \mathbb{R}$ *as* $a \to \infty$.

This result makes possible uniform approximations of the operating characteristics of sequential tests.

It is possible to refine the tangent approximation by higher order terms. Ferebee (1983) and Jennen (1985) derived those. The subsequent result, which is a bit stronger than Theorem 1 of Jennen (1985), gives a second-order approximation which holds uniformly on intervals. We need the following assumptions:

(I') $P(T_a < h_a) \to 0$ as $a \to \infty$,

(II') there exists a constant $0 < \alpha < 1$, such that $\psi_a(t)/t^\alpha$ is monotone decreasing on $\mathbb{R}_+$ for each a,

(III') ψ_a is twice continuously differentiable on $I_a = (0, h_a)$ and for every $\varepsilon > 0$ there exists a $\delta > 0$ such that for all a holds, if $s, t \in I_a$ with $|s/t - 1| < \delta$ then $|\psi_a'(s)/\psi_a'(t) - 1| < \varepsilon$ and $|\psi_a''(s)/\psi_a''(t) - 1| < \varepsilon$,

(IV') there exist constants $\rho < 1$ and $B < \infty$ such that $|t^{3/2}\psi_a''(t)| < B(\psi_a(t)/\sqrt{t})^{1+\rho}$ for all $t \in I_a$ and all a.

Theorem 4.4: _Assume the conditions_ (I')-(IV'). _Then_

$$(4.8) \quad p_a(t) = \left[\Lambda_a(t) + \frac{\psi_a''(t) t^3}{2\Lambda_a(t)^2} (1 + o(1)) - \Lambda_a(t) P(T_a < t)(1 + o(1)) \right.$$

$$\left. + o(R_a(t)) \right] t^{-3/2} \phi\left(\frac{\psi_a(t)}{\sqrt{t}}\right)$$

holds uniformly on $(0, h_a)$ _as_ $a \to \infty$. _The remainder satisfies_ $R_a(t) = \exp(-(\psi_a(t)/\sqrt{t})^\kappa)$ _for some_ $\kappa > 0$.

Concerning the proof see the remarks at the end of the section. This second order approximation consists of two correction terms, one local and one global. While the local term takes the curvature of ψ_a near t into account, the global term takes care of paths which cross early and then at t again. In the next section we shall discuss the appearance of such a term more thoroughly. Integration of (4.8) yields a result quite similar to Corollary 4.1. From this and Theorem 4.4 the following examples are obtained (cf. Jennen (1985)).

Example 1s: $\psi_a(t) = \sqrt{2(r + at)}$, $r > 0$, for $t > 0$.

$$P(T < t) = \phi(\sqrt{2a}) \left[(\sqrt{2a} - 1/\sqrt{2a} (1 + o(1))) \int_{\sqrt{2/t}}^{\infty} e^{-rx^2/2} dx/x \right.$$

$$\left. + (3/\sqrt{8a}) e^{-r/t}(1 + o(1)) \right]$$

uniformly for all $1 \le t \le \exp(a^{-p}e^a)$ with $p > 3/2$ (cf. equation (30) of Siegmund (1977)).

Example 2s: $\psi_a(t) = \sqrt{t \log \frac{a^2}{t}}$. Here we consider the one-sided case of the preceding Example 2. On $(0, t_1)$ holds

$$(4.9) \quad P(T_a \leqq t) = \frac{1}{a\sqrt{2\pi}} \left[\psi_a(t) + t/\psi_a(t)(1+o(1)) \right] \quad \text{and}$$

$$p_a(t) = \left[\psi_a(t)/2 + \frac{t}{2\psi_a(t)} \left[1 - \frac{1 + t^2/\psi_a(t)^4}{1 + t/\psi_a(t)^2} \right] \right.$$

$$\left. + o(\psi_a(t)^{-1}) \right] t^{-3/2} \phi(\frac{\psi_a(t)}{\sqrt{t}}) .$$

Finally we discuss the quality of the approximations in the case of Example 2 (two-sided case) for which the first exit density is exactly known. We compare the exact density for the upper branch

$$p_{a+}(t) = \frac{\psi_a(t)}{2t^{3/2}} \phi(\frac{\psi_a(t)}{\sqrt{t}})$$

with the tangent approximation p_1 given by (4.7) and with the second-order approximation p_2 given by (4.9). The global term is left out since it is exponentially small. Let

$$r_i = \sup_{0 < t \leqq 1} \frac{p_i(t) - p(t)}{p(t)} .$$

Then the following table shows the quality of the approximations:

a	2	5	10	20
r_1	0.5	0.2	0.1	0.05
r_2	0.222	0.055	0.017	0.005

We recognize that the second order term improves the approximation considerably. In this example presumably the effect of double crossing makes worse the quality of the approximation. The results of Jennen (1985) for a one-sided example seem to be better although we have to be careful with comparisons. Jennen calculates the true density by solving the integral equation (3.28) numerically which also leads to small errors.

The approximations of the probabilities $P(T_a<1)$, given by the formula (4.9), are contained in the following table

a	2	5	10	20	100
$P(T_a<1)$	0.404	0.188	0.104	0.057	0.013

Proof of Theorem 4.1: The following proof combines the ideas of Section 3 with a construction of the proof of Theorem 3.5 of Strassen (1967). This leads to a relative simple proof of the result, which can be used to give a higher order approximation. We begin with a lemma.

Lemma 4.1: Assumption (I) *implies*

$$(4.10) \qquad \inf_{0<t\le t_1} \frac{\psi_a(t)}{\sqrt{t}} \to \infty \qquad as\ a \to \infty.$$

The proof is the same as that of statement (3.9) of Lemma 3.2.

From now on we skip the index a. Let $\varepsilon>0$ be chosen such that $\alpha+\varepsilon<1$. Let $s=t(1-(\frac{t}{\psi(t)^2})^\varepsilon)$. Let β and γ be chosen such that

$$(4.11) \qquad \alpha+\varepsilon<2\beta-1<\gamma<\beta<1 .$$

Let $r(s)$ be the solution of the equation

$$(4.12) \qquad \frac{\psi(s)^2}{s} (\frac{r}{s})^\gamma = 1.$$

From statement (4.10) and the definitions of β and γ it follows:

$$(4.13) \qquad r/s \to 0 \quad \text{uniformly on} \quad I=(0,t_1] \quad \text{as } a \to \infty,$$

(4.14) $\dfrac{\psi(s)^2}{s}$ $(\dfrac{r}{s})^\beta$ → 0 uniformly on I as a → ∞.

(4.15) $\dfrac{\psi(s)^2}{s}$ $(\dfrac{r}{s})^{2\beta-1}$ → ∞   uniformly on I as a → ∞.

Let $p^{(x,r)}(t)$ denote the first exit density over ψ of the Brownian motion which starts at (x,r) with $x<\psi(r)$. Then by conditioning we get

(4.16) $p(t) = \displaystyle\int_{-\infty}^{\psi(r)} P(T>r,W(r)\in dx)p^{(x,r)}(t).$

We estimate this equation with the help of the subsequent lemmas.

Lemma 4.2: Let $k(r)=\psi(s)(\dfrac{r}{s})^\beta$. *Then*

(4.17) $p^{(x,r)}(T\leq s\,|\,W(t)=\psi(t))=o(1)$

uniformly for all $t\in(0,t_1)$ *and all* $x\leq k(r)$.

Proof: Let $y(s)$ be chosen such that $y(s)<\psi(s)$. Then

(4.18) $p^{(x,r)}(T\leq s\,|\,W(t)=\psi(t))$

   $= p^{(x,r)}(W(s)\geq y(s)\,|\,W(t)=\psi(t))$

   $+ p^{(x,r)}(T\leq s,W(s)<y(s)\,|\,W(t)=\psi(t))$

   $= 1-\phi(\sqrt{\dfrac{t-r}{(s-r)(t-s)}}(y(s)-\dfrac{s-r}{t-r}\psi(t)-\dfrac{t-s}{t-r}x))$

   $+ \displaystyle\int_{-\infty}^{y(s)} p^{(x,r)}(T\leq s\,|\,W(s)=y)P(W(s)\in dy\,|\,W(t)=\psi(t))$

   $= I+II.$

Let $\delta > 0$ be selected such that $0 < 1 - \delta - \epsilon$. We choose $y(s) = \frac{s}{t} \psi(t) + K_s \sqrt{\frac{s(t-s)}{t}}$ with $K_s = (\frac{\psi(s)}{\sqrt{s}})^{1-\delta-\epsilon}$. Then $y(s) < \psi(s)$. To see this we note first that by the monotonicity of ψ and by assumption (II)

(4.19) $\quad (\frac{s}{t})^{\alpha} \psi(t) \leq \psi(s) \leq \psi(t)$.

Then by (4.19) for sufficiently large a holds

$$(4.20) \quad \psi(s) - y(s) = \psi(s) - \frac{s}{t} \psi(t) - K_s \sqrt{\frac{s(t-s)}{t}}$$

$$\geq \psi(s) \left[(1 - (\frac{s}{t})^{1-\alpha}) - \frac{t-s}{t} (\frac{\sqrt{s}}{\psi(s)})^{\delta} (1 + o(1)) \right]$$

$$\geq \psi(s) \frac{1-\alpha}{2} \frac{t-s}{t}$$

$$= \frac{1-\alpha}{2} \psi(s) (\frac{t}{\psi(t)^2})^{\epsilon}.$$

Then by $r = o(t)$

$$(4.21) \quad \sqrt{\frac{t-r}{(s-r)(t-r)}} (y(s) - \frac{s-r}{t-r} \psi(t) - \frac{t-s}{t-r} k) \geq \left[K_s - \sqrt{\frac{t-s}{st}} k \right] (1 + o(1))$$

$$= \left[(\frac{\psi(s)}{\sqrt{s}})^{1-\delta-\epsilon} - \frac{k}{\sqrt{s}} (\frac{\sqrt{t}}{\psi(t)})^{\epsilon} \right] (1 + o(1))$$

$$= (\frac{\psi(s)}{\sqrt{s}})^{1-\delta-\epsilon} (1 + o(1)) \to \infty$$

since $\frac{k}{\sqrt{s}} = \frac{\psi(s)}{\sqrt{s}} (\frac{r}{s})^{\beta} \to 0$. This implies $I \to 0$.

Let g be the linear function with $g(s) = \psi(s)$ and $g(r) = \psi(s) (\frac{r}{s})^{\alpha}$. Then $g \leq \psi$ on $[r, s]$ since $h(u) = (\frac{u}{t})^{\alpha} \psi(s)$ is a concave minorant of ψ on (r, s) with $g(r) = h(r)$ and $g(s) = h(s)$.

Let $\bar{T}=\inf\{u>r\,|\,W(u)\geq g(u)\}$. Then by Example 1 of Section 1 for $y<\psi(s)$ and $x<k$

$$(4.22)\quad P^{(x,r)}(T\leq s\,|\,W(s)=y) \leq P^{(x,r)}(\bar{T}\leq s\,|\,W(s)=y)$$

$$= \exp\left(-\frac{2(g(s)-y)}{s-r}(g(r)-x)\right)$$

$$\leq \exp\left(-\frac{2}{s}(\psi(s)-y(s))(\psi(s)(\tfrac{r}{s})^{\alpha}-k)\right)$$

Since $\alpha<\beta$ and $r/s\to 0$ we get for sufficiently large a

$$(4.23)\quad \psi(s)(\tfrac{r}{s})^{\alpha}-k = \psi(s)\left[(\tfrac{r}{s})^{\alpha}-(\tfrac{r}{s})^{\beta}\right]\geq\tfrac{1}{2}\psi(s)(\tfrac{r}{s})^{\alpha}\ .$$

Since $\alpha<2\beta-1-\varepsilon$ and $\gamma<1$ we get for large a

$$(\tfrac{r}{s})^{\alpha}\geq(\tfrac{r}{s})^{2\beta-1}(\tfrac{s}{r})^{\varepsilon}\quad\text{and}\quad (\tfrac{s}{r})^{\varepsilon}=(\tfrac{\psi(s)^{2}}{s})^{\varepsilon/\gamma}\geq(\tfrac{\psi(t)^{2}}{s})^{\varepsilon/\gamma}(\tfrac{s}{t})^{2\alpha})^{\varepsilon/\gamma}$$

$$\geq(\tfrac{\psi(t)^{2}}{t})^{\varepsilon}\ .$$

This together with (4.23) leads to

$$(4.24)\quad \psi(s)(\tfrac{r}{s})^{\alpha}-k \geq \tfrac{1}{2}\psi(s)(\tfrac{r}{s})^{2\beta-1}(\tfrac{\psi(t)^{2}}{t})^{\varepsilon}.$$

Combining (4.20), (4.22) and (4.24) yields

$$(4.25)\quad P^{(x,r)}(T\leq s\,|\,W(s)=y) \leq \exp\left(-\frac{1-\alpha}{2}\frac{\psi(s)^{2}}{s}(\tfrac{r}{s})^{2\beta-1}\right).$$

From (4.25) and (4.18) by (4.15), II $\to$ 0. This proves the lemma.

□□□

Lemma 4.3: For all **x≤k(r)** *holds uniformly*

(4.26) $p^{(x,r)}(t) = \frac{\Lambda(t)(1+o(1))-x}{(t-r)^{3/2}} \phi(\frac{\psi(t)-x}{\sqrt{t-r}})$.

Proof: Let x≤k(r). Since ψ is monotone increasing, by equation (3.34) holds

(4.27) $p^{(x,r)}(t) = \frac{\psi(t)-x}{(t-r)^{3/2}} \phi(\frac{\psi(t)-x}{\sqrt{t-r}})$

$-\int_r^t p^{(x,r)}(u) \frac{\psi(t)-\psi(u)}{(t-u)^{3/2}} \phi(\frac{\psi(t)-\psi(u)}{\sqrt{t-u}}) du$.

We split the second integral up into two parts:

$\int_r^t = \int_r^s + \int_s^t = I + II$.

$II = \int_s^t p^{(x,r)}(u) \frac{\psi(t)-\psi(u)}{(t-u)^{3/2}} \phi(\frac{\psi(t)-\psi(u)}{\sqrt{t-u}}) du$

$= \psi'(t)(1+o(1))\int_s^t p^{(x,r)}(u) \frac{1}{(t-u)^{1/2}} \phi(\frac{\psi(t)-\psi(u)}{\sqrt{t-u}}) du$

by assumption (III).

Then by (3.27) and by Lemma 4.2

(4.28) $I = \frac{\psi'(t)}{\sqrt{t-r}}(1+o(1))\phi(\frac{\psi(t)-x}{\sqrt{t-r}})p^{(x,r)}(s<T\leq t|W(t)=\psi(t))$

$= \frac{\psi'(t)}{\sqrt{t-r}} \phi(\frac{\psi(t)-x}{\sqrt{t-r}})(1+o(1))$.

To estimate the first term we note that by assumption (II)

(4.29) $\dfrac{\psi(t)-\psi(u)}{t-u} \le \dfrac{\psi(t)}{t}\, \dfrac{1-(\frac{u}{t})^{\alpha}}{1-(\frac{u}{t})} \le \dfrac{\psi(t)}{t}$.

Thus by (4.29) and Lemma 4.2

$$(4.30)\quad I = \int_r^s p^{(x,r)}(u)\dfrac{\psi(t)-\psi(u)}{(t-u)^{3/2}}\,\phi\!\left(\dfrac{\psi(t)-\psi(u)}{\sqrt{t-u}}\right)du$$

$$\le \dfrac{\psi(t)/t}{\sqrt{t-r}}\,\phi\!\left(\dfrac{\psi(t)-x}{\sqrt{t-r}}\right)P^{(x,r)}(r<T\le s\,|\,W(t)=\psi(t)).$$

$$= \dfrac{\psi(t)/t}{\sqrt{t-r}}\,\phi\!\left(\dfrac{\psi(t)-x}{\sqrt{t-r}}\right)o(1).$$

Since $r/t \to 0$ (4.27), (4.28) and (4.30) yield

$$p^{(x,r)}(t) = \left[\dfrac{\psi(t)(1+o(1))-x}{(t-r)^{3/2}} - \dfrac{\psi'(t)(1+o(1))}{\sqrt{t-r}}\right]\phi\!\left(\dfrac{\psi(t)-x}{\sqrt{t-r}}\right)$$

from which the lemma follows, since by assumption (II),
$\psi'(t) < \alpha\psi(t)/t$.

□□□

Now we show

$$(4.31)\quad p(t) = \dfrac{\Lambda(t)}{t^{3/2}}\,\phi\!\left(\dfrac{\psi(t)}{\sqrt{t}}\right)(1+o(1))\ .$$

Since $k\le\psi(s)(\frac{r}{s})^{\alpha}\le\psi(r)$ we get from (4.16) and Lemma 4.3

$$(4.32)\quad p(t)\ge \int_{-k}^{k} P(T>r,W(r)\varepsilon dx)p^{(x,r)}(t)$$

$$\ge \int_{-k}^{k} P(T>r,W(r)\varepsilon dx)\,\dfrac{\Lambda(t)(1+o(1))-k}{(t-r)^{3/2}}\,\phi\!\left(\dfrac{\psi(t)+k}{\sqrt{t-r}}\right).$$

By assumption (II) and the definition of k, k=o(Λ(t)). By (4.13) and (4.14)

$$(4.33) \quad \phi(\frac{\psi(t)+k}{\sqrt{t-r}}) = \phi(\frac{\psi(t)}{\sqrt{t}})(1+o(1)) \ .$$

This together with (4.32) yields

$$(4.34) \quad p(t) \geq P(T>r,|W(r)|\leq k) \frac{\Lambda(t)}{t^{3/2}} \phi(\frac{\psi(t)}{\sqrt{t}})(1+o(1)).$$

But $P(T>r,|W(r)|\leq k) \to 1$ by assumption (I) and since
$k^2/r = \frac{\psi(s)^2}{s}(\frac{r}{s})^{2\beta-1} \to \infty$ by (4.15). Thus (4.34) yields

$$(4.35) \quad p(t) \geq \frac{\Lambda(t)}{t^{3/2}} \phi(\frac{\psi(t)}{\sqrt{t}})(1+o(1)) \ .$$

To show (4.31) by equation (4.16) it is only left to show that

$$(4.36) \quad \int_{-\infty}^{-k} P(T\geq r,W(r)\in dx)p^{(x,r)}(t)$$

$$+ \int_{k}^{\psi(r)} P(T>r,W(r)\in dx)p^{(x,r)}(t) = o(\frac{\Lambda(t)}{t^{3/2}}\phi(\frac{\psi(t)}{\sqrt{t}})) \ .$$

For the first term this is easy to see by Lemma 4.3 and since $k/\sqrt{r} \to \infty$. The estimate for the second term can be done as follows

$$\int_{k}^{\psi(r)} P(T>r,W(r)\in dx)p^{(x,r)}(t)$$

$$\leq \int_{k}^{\psi(r)} P(W(r)\in dx) \frac{\psi(t)-x}{(t-r)^{3/2}} \phi(\frac{\psi(t)-x}{\sqrt{t-r}})$$

by the monotonicity of ψ,

$$\leq \frac{\psi(t)-k}{t^{3/2}} \; \phi(\frac{\psi(t)}{\sqrt{t}}) \; (1+o(1)) \int\limits_{k}^{\psi(r)} \sqrt{\frac{t}{r(t-r)}} \; \phi(\sqrt{\frac{t}{r(t-r)}} \; (x-\frac{r}{t}\psi(t))) dx$$

$$\leq \frac{\psi(t)}{t^{3/2}} \; \phi(\frac{\psi(t)}{\sqrt{t}}) \; (1+o(1)) \left[1-\Phi\sqrt{\frac{t}{r(t-r)}} \; (k-\frac{r}{t}\psi(t)))\right]$$

$$= (1-\alpha)^{-1} \frac{\Lambda(t)}{t^{3/2}} \; \phi(\frac{\psi(t)}{\sqrt{t}}) o(1)$$

by assumption (II) since $(1-\alpha)\psi(t) \leq \Lambda(t)$ and since

$$\frac{1}{\sqrt{r}}(k-\frac{r}{t}\psi(t)) = \frac{k}{\sqrt{r}}(1-(\frac{r}{t})^{1-\beta}(1+o(1))) = \frac{k}{\sqrt{r}}(1+o(1)) \; ,$$

where $k^2/r = \frac{\psi(s)^2}{s}(\frac{r}{s})^{2\beta-1} \to \infty$ by (4.15). This yields (4.36) which completes the proof.

□□□

For the proof of a result which is a little bit weaker than Theorem 4.4 see Jennen (1985). She additionally assumes that $\psi(t)/t^{\beta}$ is monotone increasing for some $\beta>0$. This assumption rules out the preceding example (1s). It can be deleted by combining the arguments of Jennen with those of the preceding proof. Then one has to define the number s(t) a bit different to get a smaller interval (s,t). This is needed for Jennen's argument. But this demands a more tricky argument in Lemma 4.2. The details will be given elsewhere.

5. Beyond the tangent approximation and back to the Kolmogorov-Petrovski-Erdös test

Let $\psi(t)$ denote an increasing and continuously differentiable function. Let $T=\inf\{t>0 \mid W(t)\geq\psi(t)\}$ denote the first exit time of the standard Brownian motion $W(t)$ over $\psi(t)$ with $T=\infty$ if the infimum is taken over the empty set. Let $P(T>0)=1$ and let $p(t)$ denote the density of the distribution of T. For boundaries ψ which grow faster than $\sqrt{t}$ as t tends to infinity like $\sqrt{2t \log \log t}$, we study the asymptotic behaviour of $p(t)$ and of the tail probabilities $P(T>t)$. From our results the Kolmogorov-Petrovski-Erdös test near infinity will be derived. Finally we discuss uniform approximations of the first exit densities on the whole real line. Let $\Lambda(t)=\psi(t)-t\psi'(t)$ denote the intercept on the space-axis of the tangent at t to the curve ψ and let $\phi(y)=\frac{1}{\sqrt{2\pi}} e^{-y^2/2}$ denote the standard normal density.

Theorem 5.1: Let $P(T>t)>0$ *for all* $t>0$. *Assume further*

(I) $\psi(t)/\sqrt{t} \to \infty$ *as* $t \to \infty$,

(II)   *there exist a constant* $\frac{1}{2}<\alpha<1$ *such that for all* $t\geq t_o>0$, $\psi(t)/t^\alpha$ *is monotone decreasing,*

(III) *for every* $\varepsilon>0$ *there exists a* $\delta>0$ *and a* $t_1>0$ *such that* $|s/t-1|<\delta$ *implies* $|\frac{\psi'(s)}{\psi'(t)} -1|<\varepsilon$ *if* $t\geq t_1$.

Then

(5.1) $p(t) = P(T>t) \dfrac{\Lambda(t)}{t^{3/2}} \phi(\dfrac{\psi(t)}{\sqrt{t}}) (1+o(1))$ *as* $t \to \infty$.

Before we give the proof of the result we discuss several aspects and consequences of it. The special case, $P(T<\infty)<1$, is already treated in Jennen-Lerche (1982).

The asymptotic density consists of two factors, one local and one global. The local factor is the tangent approximation. An intuitive way to interpretate the global factor is this: one can think of the paths to be killed when they hit the boundary ψ. Then the theorem states that the hazard or mortality rate $p(t)/P(T>t)$ is asymptotically equal to the tangent approximation. This aspect is essential for the subsequent calculation of the tails of first exit time distributions for lower class functions. For those $P(T<\infty)=1$ holds. Since $\frac{d}{du} \log P(T>u) = -\frac{p(u)}{P(T>u)}$, we get

$$(5.2) \quad P(T>t) = \exp\left(-\int_o^t p(u)/P(T>u)\,du\right) .$$

Since $P(T>t) \to 0$ as $t \to \infty$, the integral $\int_o^t p(u)/P(T>u)\,du \to \infty$. Thus the combination of (5.1) and (5.2) yields the following corollary.

Corollary 5.1: Let $\psi(t)/\sqrt{t}$ be decreasing on an interval $(0,\varepsilon)$ and assume (I)-(III) and $P(T<\infty)=1$. Then

$$(5.3) \quad P(T>t) = \exp\left(-\int_o^t \frac{\Lambda(u)}{u^{3/2}} \phi\left(\frac{\psi(u)}{\sqrt{u}}\right) du\,(1+o(1))\right).$$

The additional condition together with the monotonicity of ψ guarantees that $\int_o^\varepsilon \frac{\psi(u)}{u^{3/2}} \phi\left(\frac{\psi(u)}{\sqrt{u}}\right) du$ is finite and thus also $\int_o^\varepsilon \frac{\Lambda(u)}{u^{3/2}} \phi\left(\frac{\psi(u)}{\sqrt{u}}\right) du$. This follows from the Kolmogorov-Petrovski-Erdös test (cf. Ito-McKean, p. 33) since $P(T>0)=1$. One can remove the additional condition of the corollary by just taking a small initial part out. For every $t_1>0$ holds

$$(5.4) \quad P(T>t) = P(T>t_1)\exp\left(-\int_{t_1}^t \frac{\Lambda(u)}{u^{3/2}} \phi\left(\frac{\psi(u)}{\sqrt{u}}\right) du\,(1+o(1))\right)$$

$$= \exp\left(-\int_{t_1}^t \frac{\Lambda(u)}{u^{3/2}} \phi\left(\frac{\psi(u)}{\sqrt{u}}\right) du\,(1+o(1))\right),$$

since the exponent tends to infinity as $t \to \infty$.

There are some related results in the literature, although they are less precise. Bass-Cranston (1983) give upper and lower bounds for $P(T>t)$, but do not get the precise asymptotic order of the exponent. Uchiyama (1980, p. 95) states an upper bound which for the boundary $(2(1-\varepsilon)t \log \log t)^{1/2}$, $\varepsilon>0$, gives the right exponent up to a constant, while for boundaries like $(2t \log \log t)^{1/2}$ gives a too large rate. Here are the exact rates for them.

Example 1: $\psi(t) = (2(1-\varepsilon)t \log \log t)^{1/2}$, $\varepsilon > 0$, for large t.

$$P(T > t) = \exp\left(-\frac{\sqrt{1-\varepsilon}}{2\sqrt{\pi \varepsilon}} (\log \log t)^{1/2} (\log t)^{\varepsilon} (1 + o(1))\right)$$

as $t \to \infty$.

Example 2: $\psi(t) = (2t \log \log t)^{1/2}$ for large t.

$$P(T > t) = \exp\left(-\frac{1}{3\sqrt{\pi}} (\log \log t)^{3/2} (1 + o(1))\right)$$

as $t \to \infty$.

Theorem 5.1 leads also to a characterization of the upper and lower class functions at infinity, a special version of the Kolmogorov-Petrovski-Erdös test. For the general form see Theorem 5.4 below.

Corollary 5.2: Let $P(T > t) > 0$ *for all* $t > 0$ *and let the assumptions* (I)-(III) *hold. Then* $P(T < \infty) < 1$ *if and only if*

$$\int_{t_1}^{\infty} \frac{\psi(t)}{t^{3/2}} \, \phi\left(\frac{\psi(t)}{\sqrt{t}}\right) dt < \infty \qquad \text{for some } 0 < t_1 < \infty \quad .$$

Proof: Since by assumption (II) and the monotonicity of ψ
$(1-\alpha)\psi(t) \leq \Lambda(t) \leq \psi(t)$ holds, the finiteness of the integral
$\int_{t_1}^{\infty} \frac{\psi(t)}{t^{3/2}} \phi\left(\frac{\psi(t)}{\sqrt{t}}\right) dt$ is equivalent to that of $\int_{t_1}^{\infty} \frac{\Lambda(t)}{t^{3/2}} \phi\left(\frac{\psi(t)}{\sqrt{t}}\right) dt$.

Let $P(T < \infty) = 1$. Then the equation (5.4) implies

$$\int_{t_1}^{\infty} \frac{\Lambda(t)}{t^{3/2}} \, \phi\left(\frac{\psi(t)}{\sqrt{t}}\right) dt = \infty.$$

Conversely if $P(T < \infty) < 1$, then $\lim_{t \to \infty} P(T > t) = P(T = \infty) > 0$. Thus by equation (5.1)
as $t' \to \infty$

$$P(t'<T<\infty) = \int_{t'}^{\infty} p(t)\,dt$$

$$= P(T=\infty)(1+o(1)) \int_{t'}^{\infty} \frac{\Lambda(t)}{t^{3/2}}\, \phi(\frac{\psi(t)}{\sqrt{t}})\,dt,$$

which implies that $\int_{t_1}^{\infty} \frac{\Lambda(t)}{t^{3/2}}\, \phi(\frac{\psi(t)}{\sqrt{t}})\,dt<\infty.$

□□□

The statement of Corollary 5.2 is unusual in the sense that it characterizes upper and lower class functions at infinity by first exit times, while the usual definition and characterizations (cf. Ito-McKean (p. 33 and p. 163) and Strassen (1967)) use last entrance times. We shall show in the proof of Theorem 5.4 that the well-known result can be derived from Corollary 5.2 under somewhat stronger conditions.

We contrast Theorem 5.1 with a result of Novikov (1981) about first exit distributions over boundaries which grow slower than parabolas, i.e. $\psi(t)=o(\sqrt{t})$ as $t \to \infty$. For those the tangent approximation no longer holds and an other phenomenon occurs. The tail probabilities are those of horizontal boundaries which are adjusted to the right level. For a proof see Novikov (1981). The following result is stated for densities.

Theorem 5.2: Let $\psi(t)$ be monotone increasing and concave. Then the integral

$$\int_{t_1}^{\infty} \frac{\psi(t)}{t^{3/2}}\,dt < \infty \quad \textit{if and only if} \quad E\psi(T)<\infty \quad \textit{and}$$

$$p(t) = \frac{c}{t^{3/2}}(1+o(1)) \quad \textit{as } t\to\infty \quad \textit{where } c = E\psi(T).$$

We note that here, since $\psi(t)/\sqrt{t}\to0$ the factor $\phi(\frac{\psi(t)}{\sqrt{t}})$, which plays such a crucial role for the high boundaries, drops out.

We also note that Uchiyama (1980) has obtained upper and lower bounds for the tail probabilities of the order $t^{-1/2}$. From Uchiyama's result the asymptotic density can be derived by using the integral equation (3.32). We omit the details.

For the proof of Theorem 5.1 we need the following two lemmas. The first one is quite intuitive. For a proof see for instance Uchiyama (1980).

Lemma 5.1: Let ψ_1 and ψ_2 denote continuous upper class functions at zero with $\psi_1 \le \psi_2$ on $(0,\tau]$. Let $T_i = \inf\{t>0 \,|\, W(t) \ge \psi_i(t)\}$. Then for $z < \psi_1(\tau)$, $\tau > 0$ holds

$$(5.5) \quad P(W(\tau) \le z \,|\, T_2 > \tau) \le P(W(\tau) \le z \,|\, T_1 > \tau).$$

With the help of this lemma the following inequality for the hazard functions can be proved. It is due to Cuzick (1981a).

Lemma 5.2: Let ψ_1 and ψ_2 denote continuous functions which are upper class at zero and which are continuous differentiable in a vicinity of t. Let $\psi_1 < \psi_2$ on $(0,t)$ and $\psi_1(t) = \psi_2(t)$. Then

$$(5.6) \quad \frac{p_1(t)}{P(T_1 > t)} \le \frac{p_2(t)}{P(T_2 > t)} \ .$$

Proof: We only sketch it. The densities p_i, i=1,2 exist by Lemma 3.3 of Strassen (1967).

$$\frac{p_1(t)\,dt}{P(T_1 > t)} = P(T_1 \in (t, t+dt) \,|\, T_1 > t)$$

$$= \int_{-\infty}^{\psi_1(t)} P(W(t) \in dy \,|\, T_1 > t)\, p^{(y,t)}(W(s) \ge \psi_1(t) \quad \text{for some s, } t < s < t+dt)$$

where $p^{(y,t)}$ denotes the measure of Brownian motion conditioned that $W(t) = y$. By the differentiability of ψ_1 only $\psi_1(t)$ enters in the integrand. This can be seen by a scaling argument. Since the integrand is monotone increasing in y, by Lemma 5.1 and the assumption $\psi_1(t) = \psi_2(t)$ we get

$$\leq \int_{-\infty}^{\psi_2(t)} P(W(t)\in dy \,|\, T_2>t)\, P^{(y,t)}(W(s)\geq\psi_2(t) \quad \text{for some } s, \ t<s<t+dt)$$

$$= \frac{p_2(t)\,dt}{P(T_2>t)} \ .$$

□□□

Proof of Theorem 5.1: We prove first

(5.7) $\dfrac{p(t)}{P(T>t)} \leq \dfrac{\Lambda(t)}{t^{3/2}} \, \phi(\dfrac{\psi(t)}{\sqrt{t}}) \, (1+o(1)) \quad \text{as } t\to\infty.$

Let $\varepsilon>0$ be chosen such that $\alpha+\varepsilon<1$. Let $s=t(1-(\dfrac{t}{\psi(t)^2})^\varepsilon)$. Let $\Lambda_1(t)= \sup\limits_{u\in[s,t]} \Lambda(u)$ and let ℓ be the linear function with $\ell(0)=\Lambda_1(t)$ and $\ell(t)=\psi(t)$. Let

$$\psi_1(u) = \begin{cases} \ell(s) & \text{for } 0<u\leq s \\[2mm] \ell(u) & \text{for } s<u\leq t \ . \end{cases}$$

Then $\psi\leq\psi_1$ on $(0,t]$.

Let $S=\inf\{u>0 \,|\, W(u)\geq\psi_1(u)\}$ and let $p_1(t)$ denote the value of the corresponding density at t. Then by Lemma 5.2

(5.8) $\dfrac{p(t)}{P(T>t)} \leq \dfrac{p_1(t)}{P(S>t)} \ .$

By Example 1 of Section 1 and assumption (I)

$$P(S>t) \geq P(W(u)<\psi(s) \quad \text{for all } 0<u\leq t)$$

$$= 2\Phi(\dfrac{\psi(s)}{\sqrt{s}}\sqrt{\dfrac{s}{t}})-1+1$$

as $s\to\infty$.

The estimate of p_1 is the same as that of Strassen (1967, p. 325):

(5.9)
$$p_1(t) = \int_{-\infty}^{\ell(s)} P(S > s, W(s) \in dx) p_1^{(x,s)}(t)$$

$$\leq \int_{-\infty}^{\ell(s)} \frac{1}{\sqrt{s}} \phi(\frac{x}{\sqrt{s}}) \frac{\ell(s)-x}{(t-s)^{3/2}} \phi(\frac{\psi(t)-x}{\sqrt{t-s}}) dx$$

$$= \frac{\Lambda_1}{t^{3/2}} \phi(\frac{\psi(t)}{\sqrt{t}}) \int_{-\infty}^{\ell(s)} (1 - \frac{t}{\Lambda_1(t-s)}(x-\frac{s}{t}\psi(t))) \sqrt{\frac{t}{s(t-s)}} \phi(\sqrt{\frac{t}{s(t-s)}}(x-\frac{s}{t}\psi(t))) dx$$

since
$$\frac{\ell(s) - \frac{s}{t}\psi(t)}{t-s} = \frac{\Lambda_1}{t} \quad ,$$

$$= \frac{\Lambda_1}{t^{3/2}} \phi(\frac{\psi(t)}{\sqrt{t}}) (1 + o(1))$$

by the assumptions (I) and (II). Since $\frac{s}{t} \to 1$ by the assumptions (II) and (III) $\Lambda_1(t)/\Lambda(t) \to 1$. Combining now (5.8) and (5.9) yields (5.7).

To prove the converse inequality of (5.7), we use the corresponding part of the proof of Theorem 4.1 up to equation (4.34). We only have to substitute the limit operations "a $\to \infty$" by "t $\to \infty$". But this does not affect the arguments. Equation (4.34) then states that

(5.10)
$$p(t) \geq P(T > r, |W(r)| \leq k) \frac{\Lambda(t)}{t^{3/2}} \phi(\frac{\psi(t)}{\sqrt{t}}) (1 + o(1))$$

where r and k are defined as in the proof of Theorem 4.1.

We show now that

(5.11)
$$P(T > r) = P(T > r, |W(r)| \leq k)(1 + o(1)).$$

(5.11) together with (5.10) yields

$$(5.12) \quad p(t) \geq P(T>r)\frac{\Lambda(t)}{t^{3/2}} \phi(\frac{\psi(t)}{\sqrt{t}})(1+o(1))$$

$$\geq P(T>t) \frac{\Lambda(t)}{t^{3/2}} \phi(\frac{\psi(t)}{\sqrt{t}})(1+o(1))$$

which then completes the proof.

It is left to show equation (5.11). This is not trivial since both sides of the equation may tend to zero. At first we prove

$$(5.13) \quad P(T>r) = P(W(r)\leq k,T>r)(1+o(1)).$$

Let $S=\inf\{u>0|W(u)\geq\psi(r)\}$. Since ψ is monotone increasing and $k<\psi(r)$ Lemma 5.1 yields

$$(5.14) \quad P(W(r)\leq k|T>r) \geq P(W(r)\leq k|S>r).$$

We calculate the right hand side by using a result of Example 1 of Section 1.

$$(5.15) \quad P(W(r)\leq k|S>r) = \frac{\phi(\frac{k}{\sqrt{r}})-\phi(\frac{k-2\psi(r)}{\sqrt{r}})}{2\phi(\frac{\psi(r)}{\sqrt{r}})-1} \ .$$

Since $\frac{\psi(r)}{\sqrt{r}} \geq \frac{k}{\sqrt{r}} \to \infty$ by (4.15) and since $(k-2\psi(r)/\sqrt{r} \to -\infty$, the right hand side of (5.15) converges to one and thus by (5.14) proves (5.13).

The next step is to show

$$(5.16) \quad P(W(r)\leq -k,T>r) = P(T>r)o(1).$$

Let $0<q<r$ be a fixed time point . Let $\psi_2(u) = \begin{cases} \psi(u) & \text{for } u\leq q \\ \psi(q) & \text{for } u>q \end{cases}.$

Let $U=\inf\{u>0\,|\,W(u)\geq\psi_2(u)\}$. Since $\psi_2(u)\leq\psi(u)$ for $u\leq r$, Lemma 5.1 yields

(5.17) $\quad P(W(r)\leq-k\,|\,T>r)\leq P(W(r)\leq-k\,|\,U>r)$

$$=\frac{\int_{-\infty}^{\psi(q)}P^{(x,q)}(W(r)\leq-k,U>r)H_q(dx)}{\int_{-\infty}^{\psi(q)}P^{(x,q)}(U>r)H_q(dx)}$$

where $P^{(x,q)}$ denotes the measure of Brownian motion conditioned that $W(q)=x$ and $H_q(dx)=P(U>q,W(q)\in dx)$.

$$\leq\frac{\int_{-\infty}^{-\sqrt{r}}\frac{1}{\sqrt{q}}\,\phi(\frac{x}{\sqrt{q}})\,dx}{\int_{-\infty}^{0}\left[2\Phi(\frac{\psi(q)-x}{\sqrt{r-q}})-1\right]H_q(dx)}$$

$$+\frac{\int_{-\sqrt{r}}^{\psi(q)}\left[\Phi(\frac{-k-x}{\sqrt{r-q}})-\Phi(\frac{-k-x-2(\psi(q)-x)}{\sqrt{r-q}})\right]H_q(dx)}{\int_{-\sqrt{r}}^{\psi(q)}\left[2\Phi(\frac{\psi(q)-x}{\sqrt{r-q}})-1\right]H_q(dx)}$$

by Example 1 of Section 1.

$$\leq\frac{1-\Phi(-\sqrt{\frac{r}{q}})}{P(U>q,W(q)<0)\left[2\Phi(\frac{\psi(q)}{\sqrt{r-q}})-1\right]}$$

$$+\frac{\int_{-\sqrt{r}}^{\psi(q)}(\int_{0}^{2x_q}\phi(y+\frac{k+x}{\sqrt{r-q}})\,dy)\,H_q(dx)}{\int_{-\sqrt{r}}^{\psi(q)}(\int_{0}^{2x_q}\phi(y)\,dy)\,H_q(dx)}$$

with $x_q=\frac{\psi(q)-x}{\sqrt{r-q}}$.

The first term converges to zero since r → ∞ and q is fixed. To see that r→∞, we note that by the definition of r and by assumption (II)

$$r = s\left(\frac{s}{\psi(s)^2}\right)^{1/\gamma} \geq K \; s^{1-(2\alpha-1)/\gamma}$$

which tends to infinity since by (4.11) $2\alpha-1<\gamma$ holds. The second term is smaller than $\exp\left(-\frac{(k-\sqrt{r})^2}{2(r-q)}\right)$ which tends to zero by (4.15). This implies that the right hand side of (5.17) converges to zero and thus also the left hand side. This yields (5.16). But (5.14) and (5.16) imply (5.11). This completes the proof.

□□□

A similar result as Theorem 5.1 holds for the Brownian motion with drift θ. Similar arguments as those preceding Theorem 4.3 yield the following corollary. Let p_θ denote the density of the distribution of T for the Brownian motion with drift θ.

Theorem 5.3: Let $P_o(T>t)>0$ _for all_ t>0 _and assume the conditions_ (I)-(III). _Then_

$$(5.18) \quad p_\theta(t) = P_o(T>t) \frac{\Lambda(t)}{t^{3/2}} \phi\left(\frac{\psi(t)-\theta t}{\sqrt{t}}\right)(1+o(1))$$

uniformly for all $\theta \in \mathbb{R}$ _as_ t→∞.

A consequence of this result is the asymptotic behaviour of the moments of T for upper class functions. Let a_θ denote the positive number which satisfies $\theta a_\theta = \psi(a_\theta)$. This is the time when the ray from the origin with slope θ crosses the curve ψ.

Corollary 5.3: Let k>0. _As_ θ↘0

$$E_\theta T^k = P_o(T=\infty) a_\theta^k (1+o(1)) \; .$$

The following considerations show that the usual Kolmogorov-Petrovski-Erdös-test under somewhat stronger conditions is a consequence of

Theorem 5.1 (cf. Ito-McKean, p. 33). For a continuous function ψ let the last entrance time of Brownian motion below ψ be given by

$$S = \sup\{s>0 \mid W(s) \geq \psi(s)\} .$$

If the supremum is taken over the empty set let $S=0$. Then ψ is defined as an upper (lower) class function at infinity if $S<\infty$ ($S=\infty$) almost surely.

Theorem 5.4: Let ψ be a positive continuous function on $\mathbb{R}_+$. Assume that $\psi(t)/\sqrt{t}$ is finally increasing and $\psi(t)/t^\alpha$ is finally decreasing for some $\frac{1}{2}<\alpha<1$. Then ψ is an upper class function at infinity if and only if

$$\int_{t_1}^\infty \frac{\psi(t)}{t^{3/2}} \, \phi\left(\frac{\psi(t)}{\sqrt{t}}\right) dt < \infty$$

for some $t_1>0$.

Proof: We assume first that ψ is continuously differentiable and that it satisfies the assumption (III) of Theorem 5.1. Without loss of generality we can also assume that assumption (I) holds which means $\psi(t)/\sqrt{t}\to\infty$. Another modification of the boundary is necessary. Without changing the properties of S, we can change the boundary on a finite interval $(0,t_o)$ such that $P(T>t)>0$ for all $t>0$ holds. Then

(5.18) $P(T<\infty) = 1$ if and only if $P(S=\infty) = 1$.

This follows from the inequality $T\leq S$ for the pathes which cross the boundary, and by the assumption "$P(T>t)>0$ for all t".

Now Corollary 5.1 together with (5.18) imply the result for smooth boundaries.

For arbitrary continuous boundaries one obtains the result by regularization just as in the proof of Corollary 3.7 of Strassen (1967).

□□□

Now we turn back to the case that the boundaries recede to infinity. The following result gives an approximation on the whole positive real axis. It states that for lower class functions the global second order term of Theorem 4.4 becomes a first order term when the approximation is uniform on $\mathbb{R}_+$.

Theorem 5.5: _Let $\{\psi_a; a \in \mathbb{R}_+\}$ be a set of monotone increasing, continuously differentiable functions. Assume that_

(I') $\psi_a(t)/\sqrt{t} \to \infty$ _uniformly on $\mathbb{R}_+$ as $a \to \infty$,_

(II')   _there exists a constant $\frac{1}{2} < \alpha < 1$ such that $\psi_a(t)/t^\alpha$ is decreasing,_

(III') _for every $\varepsilon > 0$ there exists a $\delta > 0$ such that for all a_

$$|\psi_a'(s)/\psi_a'(t) - 1| < \varepsilon \quad if \quad |s/t - 1| < \delta$$

for $s, t \in \mathbb{R}_+$,

(IV')   _there exists a $\gamma > 0$ such that $P(T_a < \gamma) \to 0$ as $a \to \infty$._

Then

$$p_a(t) = P(T_a > t) \; \frac{\Lambda_a(t)}{t^{3/2}} \; \phi\left(\frac{\psi_a(t)}{\sqrt{t}}\right)(1 + o(1))$$

uniformly on $\mathbb{R}_+$ as $a \to \infty$.

The proof of Theorem 5.5 follows by combining the arguments of Theorem 4.1 with those of Theorem 5.1.

With our method of proof it is possible to refine the statement of Theorem 5.5 by a second order approximation, which improves Jennen's result, Theorem 4.4, in the same way as Theorem 5.5 does the tangent approximation. The details will be discussed elsewhere. We close this section with giving necessary and sufficient conditions for the tangent approximation.

Corollary 5.4: *Let the assumptions* (I')-(IV') *hold. Let* $\{h_a ; a \in \mathbb{R}_+\}$ *denote a function with* $\lim\limits_{a\to\infty} h_a = \infty$. *Then*

$$p_a(t) = \frac{\Lambda_a(t)}{t^{3/2}} \phi(\frac{\psi_a(t)}{\sqrt{t}}) (1+o(1))$$

uniformly on $(0, h_a)$ *as* $a \to \infty$ *if and only if* $P(T_a < h_a) \to 0$.

Supplement to Chapter I

Supplement: The tangent approximation is a formal saddlepoint approximation.

The theory about the tangent approximation has been presented in the preceding chapter as an extension of the fluctuation theory of Brownian motion. There is another view possible which appeared so far mainly (somewhat hidden) in the proofs. It is from the large deviation standpoint. To describe it we need some well-known facts about the saddlepoint approximation. For a detailed discussion of this topic see Daniels (1954). Here we proceed non-rigorously.

Let $X_1, X_2, \ldots, X_n$ be independent identically distributed random variables with probability density $p_1(x)$ and with finite moment generating function $M(\theta) = \int e^{\theta x} p_1(x) dx$ in a vicinity of $\theta = 0$. Let $K(\theta) = \log M(\theta)$. $K(\theta)$ is infinitely often differentiable and convex. Let $S_n = \sum_{i=1}^{n} X_i$, $\overline{X}_n = S_n/n$ and p_n be the density of the distribution of $\overline{X}_n$. The saddlepoint approximation of $p_n(x)$ is given by

$$(S.1) \quad \tilde{p}_n(x) = \sqrt{\frac{n}{2\pi K''(\hat{\theta})}} \; \exp\left[n\left(K(\hat{\theta}) - x\hat{\theta}\right)\right]$$

where $\hat{\theta}$ is defined by $K'(\hat{\theta}) = x$.

For many distributions

$$(S.2) \quad p_n(x) = \tilde{p}_n(x)(1+o(1))$$

holds uniformly in x when $n \to \infty$ (see Daniels (1954)).

$\tilde{p}_n$ can also be expressed by the Legendre-transform of K, the entropy function. It is given by $H(x) = \sup_{\theta} \{\theta x - K(\theta)\} = \hat{\theta} K'(\hat{\theta}) - K(\hat{\theta})$. Then

$$(S.3) \quad \tilde{p}_n(x) = \sqrt{\frac{n H''(x)}{2\pi}} \; \exp(-nH(x)).$$

The saddlepoint approximation unifies the central limit theorem, the law of the iterated logarithm and the law of large numbers in a distributional sense, since these laws are consequences of (S.2). Of course, the usual large deviation statement also follows from (S.2). Let $\mu = \int xp(x)\,dx = EX_1$. Let $A \subset \mathbb{R}$ with $(\mu - \varepsilon, \mu + \varepsilon) \subset A^c$. Then by Laplace's method

$$\lim_{n \to \infty} n^{-1} \log P(\bar{X}_n \in A) = -\inf_{x \in A} H(x).$$

For later purposes we express $\tilde{p}_n$ in a form with a more statistical interpretation. For all parameters θ with $M(\theta) < \infty$ let $P_{\theta,n}(dx) = \exp(n(\theta x - K(\theta)))p_n(x)\,dx$ denote the measures of the exponential family generated by p_n. Then $\hat{\theta}$, defined by the equation $K'(\hat{\theta}) = x$, is the maximum likelihood estimator of θ, and

(S.4) $$\tilde{p}_n(x) = \sqrt{\frac{nH''(x)}{2\pi}} \left[\frac{dP_{\hat{\theta},n}}{dP_{0,n}} \right]^{-1} \quad \text{holds.}$$

So far we considered the saddlepoint approximation of the density of the (random) drift $(\bar{X}_n)$ for a fixed sample size n (which of course increases to infinity). The question arises whether similar approximations hold for stopping times $\{T_a; a > 0\}$ which increase to infinity when $a \to \infty$. Such a situation has some additional features. The history of the paths will play a role, overshoot effects can occur and the Laplace transform of $\bar{X}_T$ for a stopping time T in general will be unknown. All this makes the analysis more difficult than in the fixed sample case and implies that the saddlepoint technique of complex analysis, which leads to the name of the approximation, cannot be applied in general. Nevertheless it can be shown that results of the type (S.2) still hold for an approximation term, which is formally of the type (S.1). We call it the formal saddlepoint approximation* and discuss it in more detail for the case of Brownian motion. (The random walk problem has been considered by Klein (1986).)

---

*H. Dinges has proposed the notion "Wiener germ" for approximations of the type (S.1), (S.2). By choosing a completely new name he wants to point out that those approximations are not necessarily linked to the saddlepoint technique of complex analysis.

Let W(t) denote standard Brownian motion. Let $\psi(t)$ denote a positive concave function on $\mathbb{R}_+$ and let T denote the stopping time $T = \inf \{t > 0 \mid W(t) > \psi(t)\}$. For a given t let

(S.5) $\mu = \psi(t)/t$.

Since ψ is concave (S.5) establishes a one-to-one correspondence between t and μ. We write t_μ (respectively μ_t) when we consider t as a function of μ (μ as a function of t).

Now we calculate the formal saddlepoint approximation for $\mu_T = \psi(T)/T$ by (formally) applying the formula (S.4). Let P_θ denote the measure of Brownian motion with drift θ. The Radon-Nikodym derivative at time T is given by

$$\frac{dP_{\theta,T}}{dP_{o,T}} = \exp\left(\theta W(T) - \frac{1}{2}\theta^2 T\right)$$

$$= \exp\left(\theta \psi(T) - \frac{1}{2}\theta^2 T\right).$$

For $T = t$ we get (with $\mu_t = \psi(t)/t$)

(S.6) $\left.\dfrac{dP_{\theta,T}}{dP_{o,T}}\right|_{T=t} = \exp\left(t\left(\theta \mu_t - \frac{1}{2}\theta^2\right)\right).$

Let $\hat{\theta}_T$ denote the maximum likelihood estimator $\hat{\theta}_T = \psi(T)/T$. Formula (S.4) and (S.6) yield

(S.7) $\tilde{p}_T(\mu) = \sqrt{\dfrac{t_\mu}{2\pi}} \left[\left.\dfrac{dP_{\hat{\theta},T}}{dP_{o,T}}\right|_{T=t_\mu}\right]^{-1}$

$$= \sqrt{\frac{t_\mu}{2\pi}} \exp\left(-\frac{t_\mu}{2}\left(\frac{\psi(t_\mu)}{t_\mu}\right)^2\right)$$

$$= \sqrt{\frac{t_\mu}{2\pi}} \, \exp(-t_\mu \, \mu^2/2).$$

This formula is formally of the type (S.2) although here sample size and drift are coupled by the equation (S.5). Of course $\tilde{p}_T$ can also be expressed by the relative entropy:

$$(S.8) \qquad \tilde{p}_T(\mu) = \sqrt{\frac{t_\mu}{2\pi}} \, \exp(-t_\mu \, H(\mu))$$

where $\quad H(\gamma) = \sup_\theta \{\theta\gamma - \frac{1}{2}\theta^2\}.$

The results of Section 4 can be translated to statements of the type (S.2). For that we show that after a change of variables the formal saddlepoint approximation becomes the tangent approximation and conversely.

Let p_T denote the density of the distribution of $\mu_T = \psi(T)/T$ and let p denote the density of the distribution of T under Brownian motion with drift zero. Then

$$p(t) = -p_T(\mu_t) \, \frac{d\mu_t}{dt} \, .$$

But $\dfrac{d\mu_t}{dt} = \dfrac{d}{dt}\left(\dfrac{\psi(t)}{t}\right) = \dfrac{-\Lambda(t)}{t^2}$ with $\Lambda(t) = \psi(t) - t\psi'(t)$. Thus

$$(S.9) \qquad p(t) = \frac{\Lambda(t)}{t^2} \, p_T\!\left(\frac{\psi(t)}{t}\right).$$

Of course the formal saddlepoint approximation transforms in the same way and becomes for $p(t)$ by (S.7) and (S.9):

$$(S.10) \qquad \tilde{p}(t) = \frac{\Lambda(t)}{t^2} \, \tilde{p}_T\left(\frac{\psi(t)}{t}\right)$$

$$= \frac{\Lambda(t)}{t^{3/2}} \cdot \phi\left(\frac{\psi(t)}{\sqrt{t}}\right) \qquad ,$$

which is just the tangent approximation.

We look now back to Section 4 and translate Theorem 4.2 to an approximation result for μ_T. Of course we use the setup of Section 4. By the formulas (S.9) and (S.10) the statement of Theorem 4.2 becomes

(S.11) $\quad p_{T_a}(\mu) = \tilde{p}_{T_a}(\mu)(1+o(1))$

uniformly on the intervals $(0, h_a/a)$ as $a \to \infty$. We leave it to the reader to translate the other results, for example those of Section 5. We finally note that statement (S.11) holds for straight lines without a $o(1)$-term. For random walks Klein (1986) has derived similar results to (S.11). Of course in his case problems with the overshoot occur.

We close this section with some background information about the organization of Chapter I. It starts with the general method of images, since the tangent approximation can be derived from it in a natural way (see Section 3). Originally we tried to derive the tangent approximation from the method of mixtures of likelihood functions, but we did not succeed in general. The basic idea for that approach consists in using the following formula, which looks quite related to the formal saddlepoint approximation:

(S.12) $\quad P_o(t_o < T < t_1) = \displaystyle\int_{\{t_o<T<t_1\}} \sqrt{\frac{T}{2\pi}}\, e^{-W(T)^2/2T} dQ$

with $Q = \int P_\mu \frac{d\mu}{\sqrt{2\pi}}$. Equation (S.12) is a direct consequence of the optional stopping theorem applied to the martingale (under Q)

$$\frac{dP_{o,t}}{dQ_t} = \left(\frac{dQ_t}{dP_{o,t}}\right)^{-1} = \left(\int \frac{dP_{\mu,t}}{dP_{o,t}} \frac{d\mu}{\sqrt{2\pi}}\right)^{-1}$$

$$= \sqrt{\frac{t}{2\pi}}\, e^{-W(t)^2/2t}$$

Since one can express equation (S.12) also as

$$P_0(t_0 < T < t_1) = \int \frac{d\mu}{\sqrt{2\pi}} \, E_\mu (\sqrt{T} \, e^{-\psi(T)^2/2T} \, 1_{\{t_0<T<t_1\}})$$

one is tempted to conclude by the strong law of large numbers, that its right hand side becomes asymptotically equal to

$$\int_{\mu_1}^{\mu_0} \sqrt{\frac{t_\mu}{2\pi}} \, e^{-\psi(t_\mu)^2/2t_\mu} \, d\mu \; ,$$

where μ_i, $i=0,1$ is given by the solution of the implicit equation $\mu_i t_i = \psi(\mu_i)$. The crucial difficulty with that idea is to show that

$$(S.13) \quad E_\mu (\sqrt{T} \, e^{-\psi(T)^2/2T} 1_{\{t_0<T<t_1\}})$$

$$= \sqrt{t_\mu} \, e^{-\psi(t_\mu)^2/2t_\mu} (1+o(1))$$

holds for $\mu \in (\mu_1,\mu_0)$ when ψ recedes to infinity. For nearly parabolic boundaries this can be done (see Lai-Siegmund (1977)) since the exponential term of the left hand side of equation (S.13) behaves nicely and is nearly constant. This is not the case for straight line boundaries or those of the form at^α, $0 < \alpha < 1/2$. For those (S.13) does not hold.

CHAPTER II

OPTIMAL PROPERTIES OF SEQUENTIAL TESTS
WITH PARABOLIC AND NEARLY PARABOLIC BOUNDARIES.

1. Bayes tests of power one

Let $W(t)$ denote Brownian motion with unknown drift $\theta \in \mathbb{R}$ and P_θ the associated measure. We consider the following sequential decision problem. Let F be a prior on $\mathbb{R}$ given by $F = \gamma \delta_0 + (1-\gamma) \int \phi(\sqrt{r}\theta) \sqrt{r} d\theta$ with $0 < \gamma < 1$ and $\phi(x) = \frac{1}{\sqrt{2\pi}} e^{-x^2/2}$, consisting of a point mass at $\{\theta=0\}$ and a smooth normal part on $\{\theta \neq 0\}$. Let the sampling costs be $c\theta^2$, with $c>0$, for the observation of W per unit time when the underlying measure is P_θ. We assume also a loss function which is equal to 1 if $\theta=0$ and we decide in favour of "$\theta \neq 0$" and which is identically 0 if $\theta \neq 0$. A statistical test consists of a stopping time T of Brownian motion where stopping means a decision in favour of "$\theta \neq 0$".

The Bayes risk for this problem is then given by

$$(1.1) \qquad \rho(T) = \gamma P_0(T<\infty) + (1-\gamma)c \int_{-\infty}^{\infty} \theta^2 E_\theta T \phi(\sqrt{r}\theta) \sqrt{r} d\theta.$$

In this section we investigate the "optimal" stopping rule T_c^* which minimizes $\rho(T)$.

For the cost c sufficiently small, T_c^* is a test of power one for the decision problem $H_0 : \theta=0$ versus $H_1 : \theta \neq 0$. This is by definition (cf. Robbins (1970)) a stopping time T which satisfies the conditions

$$(1.2) \qquad P_0(T<\infty) < 1$$

$$(1.3) \qquad P_\theta(T<\infty) = 1 \qquad \text{if } \theta \neq 0.$$

A typical example of a test of power one is given by

$$(1.4) \qquad T = \inf\{t>0 \mid |W(t)| \geq \sqrt{(t+r)(\log(\frac{t+r}{r})+2\log b)}\} \text{ with } b>1.$$

For it holds $P_0(T < \infty) = b^{-1}$ as is shown by Robbins-Siegmund (1970) and by Theorem 2.1 of Chapter I.

The following rough geometric argument indicates that tests with nearly parabolic boundaries will be good procedures for the risk (1.1). For the problem of simple hypotheses given by

(1.5) $\quad \rho(T) = \gamma P_o(T < \infty) + (1 - \gamma) c \theta^2 E_\theta T$,

we know from the introduction that the optimal stopping rule is equal to

(1.6) $\quad T_c^* = \inf\{t > 0 \mid W(t) \geq \log a(\gamma, c) / \theta + \frac{1}{2}\theta t\}$

with $a(\gamma, c) = \gamma(2(1 - \gamma)c)^{-1}$ when $a(\gamma, c) > 1$ and $T_c^* = 0$ otherwise. This means that the optimal boundary is given by a straight line with slope $\theta/2$.

One gets an idea about the optimal boundary for the risk (1.1) by varying the solution of (1.5) over the different drifts θ. The lower envelope of the optimal straight lines $\log a(\gamma, c) / \theta + \frac{1}{2}\theta t$ is given by the parabola $[2 \log a(\gamma, c) t]^{1/2}$. Unfortunately this parabola does not define a test of power one. There is, however, a grain of truth in this argument, which becomes apparent by a refined heuristic consideration given in Section 2 and by the following exact result, which is proved in Section 4.

Let T_c^* denote the optimal stopping rule for the risk (1.1). Its existence can be shown by backward induction.

Let $F_{x,t}$ denote the posterior distribution with respect to the prior F given that the process $(W(u), u)$ has reached the space-time point (x, t). Let $N(\mu, \sigma^2)$ denote the normal distribution with mean μ and variance σ^2. For the prior $F = \gamma \delta_o + (1 - \gamma) N(0, r^{-1})$ the posterior is given by

(1.7) $\quad F_{x,t} = \gamma(x,t) \delta_o + (1 - \gamma(x,t)) N(\frac{x}{t+r}, \frac{1}{t+r})$

where $\quad \gamma(x,t) = \dfrac{\gamma dP_{o,t}(x)}{\gamma dP_{o,t}(x) + (1 - \gamma) \int_{-\infty}^{\infty} dP_{\theta,t}(x) N(0, r^{-1})(d\theta)}$

$$= \frac{\gamma}{\gamma + (1-\gamma) \int_{-\infty}^{\infty} \frac{dP_{\theta,t}}{dP_{o,t}} (x) \phi(\sqrt{r}\theta) \sqrt{r} d\theta}$$

with

$$(1.8) \quad \int_{-\infty}^{\infty} \frac{dP_{\theta,t}}{dP_{o,t}} (x) \phi(\sqrt{r}\theta) \sqrt{r} d\theta = \sqrt{\frac{r}{t+r}} \exp\left(\frac{x^2}{2(t+r)}\right) .$$

We note that $N(\frac{x}{t+r}, \frac{1}{t+r})$ is the posterior at (x,t) with respect to the prior $N(0, r^{-1})$. Therefore the formula (1.7) expresses the following fact: the conditional posterior distribution of θ, given that $\theta \neq 0$, is the posterior with respect to the conditional prior given that $\theta \neq 0$. The same for the conditional posterior given that $\theta = 0$ holds. Thus the posterior $F_{x,t}$ transforms in canonical way with respect to the singular and absolute continuous parts of the prior.

Let $T_\lambda = \inf\{t>0 \mid F_{W(t),t}\{0\} \leq \lambda\}$.
This stopping time defines a simple Bayes rule for a given $0<\lambda<1$. Obviously it can also be expressed as

$$T_\lambda = \inf\{t>0 \mid \int_{-\infty}^{\infty} \frac{dP_{\theta,t}}{dP_{o,t}} \phi(\sqrt{r}\theta) \sqrt{r} d\theta \geq b(\gamma,\lambda)\} ,$$

with $b(\gamma,\lambda) = \frac{(1-\lambda)\gamma}{\lambda(1-\gamma)}$. A simple calculation using (1.8) shows that T_λ coincides with the test given by (1.4):

$$T_\lambda = \inf\{t>0 \mid |W(t)| \geq \sqrt{(t+r)(\log(\frac{t+r}{r}) + 2\log b(\gamma,\lambda))}\} .$$

The following theorem states that the optimal stopping rule T_c^* is bounded from above and below by simple Bayes rules. The details of the proof of Theorem 1.1 are discussed in Section 4. Also a refinement of the result is given there.

Theorem 1.1: There exists a constant M>2 such that for every c>0

$$(1.9) \quad T_{Mc} \leq T_c^* \leq T_c \qquad holds.$$

We draw one conclusion about the asymptotic shape of the boundary of T^*_c from Theorem 1.1. Let $\psi^*_{1/c}(t)$ denote the optimal stopping boundary. It is intuitively clear that by symmetry T^*_c can be expressed as

$$T^*_c = \inf\{t>0 \mid |W(t)| \geq \psi^*_{1/c}(t)\}.$$

Thus (1.4) and (1.9) yield the following result.

Corollary 1.1: For c *fixed*

(1.10) $\psi^*_{1/c}(t) = (t \log t)^{1/2}(1+o(1))$ *as* $t \to \infty$.

Statement (1.10) was the first result of our study. We derived it at first heuristically from the tangent-approximation. This approach is discussed in the next section.

2. An application of the tangent approximation: a heuristic derivation of the shape of Bayes tests of power one

This section describes our original approach to the problem of calculating the shape of Bayes tests of power one.

For the first exit densities of one-sided boundaries ψ_a which satisfy the conditions of Theorem 4.1 of Chapter I, holds $p_a(t) = q_a(t)(1+o(1))$ uniformly on $\mathbb{R}_+$ when $a \to \infty$. Here

$$q(\psi_a(t),t) = \frac{\Lambda_a(t)}{t^{3/2}} \phi\left(\frac{\psi_a(t)}{\sqrt{t}}\right) \quad \text{with} \quad \Lambda_a(t) = \psi_a(t) - t\psi_a'(t).$$

We apply the tangent-approximation to the Bayes problem according to the following program:

1) First we rewrite the Bayes risk

$$\rho(T) = \gamma P_0(T < \infty) + (1-\gamma)c \int \theta^2 E_\theta T \phi(\sqrt{T}\theta)\sqrt{T}\, d\theta$$

as an integral over the Brownian motion without drift:
$\rho(T) = E_0 k(|W(T)|,T).$

2) For the relevant competitors for optimality $T = \inf\{t > 0 \mid |W(t)| \geq \psi(t)\}$ we use the tangent approximation for the distribution of T, $(p(t) \cong 2q(\psi(t),t))$ and rewrite the Bayes risk (approximately) as an integral over time

$$\rho(T) = E_0 k(|W(T)|,T) \cong 2\int_0^\infty k(\psi(s),s)q(\psi(s),s)\,ds.$$

3) Then we vary ψ and compute the optimal solution asymptotically when $t \to \infty$.

For simplicity we go through this program with $\gamma = \frac{1}{2}$. We show at first

(2.1) $\rho(T) = \frac{1}{2} \int\limits_{\{T<\infty\}} (1 + c\sqrt{r} \, \frac{W(T)^2}{(T+r)^{3/2}} \exp(\frac{W(T)^2}{2(T+r)})) dP_o$.

Let $\bar{Q} = \int\limits_{-\infty}^{\infty} P_\theta \phi(\sqrt{r}\theta)\sqrt{r}d\theta$. Then

$$P_{\theta,t}(dW)\phi(\sqrt{r}\theta)\sqrt{r}d\theta = G_{W(t),t}(d\theta)\bar{Q}(dW)$$

with $G_{x,t} = N(\frac{x}{t+r}, \frac{1}{t+r})$ where $N(\mu,\sigma^2)$ denotes the normal distribution with mean value μ and variance σ^2. By Fubini's theorem we get

$$\int\limits_{-\infty}^{\infty} \theta^2 E_\theta T\phi(\sqrt{r}\theta)\sqrt{r}d\theta = \int\limits_{-\infty}^{\infty} \theta^2 (\int (T+r)dP_\theta)\phi(\sqrt{r}\theta)\sqrt{r}d\theta - 1$$

$$= \int (T+r)(\int\limits_{-\infty}^{\infty} \theta^2 N(\frac{W(T)}{T+r}, \frac{1}{T+r})(d\theta))d\bar{Q} - 1$$

$$= \int (T+r)((\frac{W(T)}{T+r})^2 + \frac{1}{T+r})d\bar{Q} - 1$$

$$= \int \frac{W(T)^2}{T+r} d\bar{Q} .$$

But by equation (1.8)

(2.2) $\dfrac{d\bar{Q}_T}{dP_{o,T}} = (\frac{r}{T+r})^{1/2} \exp(\frac{W(T)^2}{2(T+r)})$.

Combining these formulas we obtain (2.1).

For the second step of our program we rewrite the risk (2.1) further. We introduce $T = \inf\{t > 0 \mid |W(t)| \geq \psi(t)\}$ with $\psi(t) = [ty(t)]^{1/2}$ and get

$$\rho(T) = \frac{1}{2} \int\limits_{\{T<\infty\}} (1 + c\sqrt{r} \, \frac{Ty(T)}{(T+r)^{3/2}} \exp(\frac{Ty(T)}{2(T+r)})) dP_o .$$

Plugging in the tangent-approximation $2\dfrac{\Lambda(t)}{t^{3/2}}\phi(\sqrt{y(t)})$ with

$\Lambda(t) = \psi(t) - t\psi'(t) = \frac{1}{2}[ty(t)]^{1/2}(1 - t\frac{y'(t)}{y(t)}))$ yields

$$\rho(T) \cong \frac{1}{2} \int_0^\infty (\frac{\sqrt{y}}{t} - \frac{y'}{\sqrt{y}}) \, (\exp(-y/2) + c\sqrt{r} \, \frac{ty}{(t+r)^{3/2}} \exp(-\frac{ry}{2(t+r)})) \frac{1}{\sqrt{\pi}} \, dt.$$

Now we try to minimize the integral on the right hand side, whose integrand we call $F(y,y',t)$. For a minimum of $\int_0^\infty F(y,y',t) \, dt$, it is necessary that the Euler-Lagrange equation $F_y - \frac{d}{dt}F_{y'} = 0$ holds (see Courant-Hilbert I, Chapter 4). By a straightforward but lengthy calculation this reduces to

$$(1-y)\exp(\frac{-yt}{2(t+r)}) + \frac{cyr^{1/2}}{(t+r)^{1/2}} \left[4\frac{t}{t+r} - \frac{3t^2+tyr}{(t+r)^2} \right] = 0 \ ,$$

from which finally $\psi(t) \sim [(t+r)(\log(\frac{t+r}{r})]^{1/2}$ follows when $t \to \infty$. This statement agrees with Theorem 1.1 and its corollary.

By playing around with equation (2.2) we can derive an exact upper bound for the optimal stopping region for the risk (1.1). This risk can be expressed as

$$(2.3) \qquad \rho(T) = \int (\gamma(\frac{T+r}{r})^{1/2} \exp(-\frac{W(T)^2}{2(T+r)}) + (1-\gamma)c \, \frac{W(T)^2}{T+r}) \, d\bar{Q}$$

$$= \int g(\frac{|W(T)|}{(T+r)^{1/2}}, T+r) \, d\bar{Q}$$

with $g(\lambda,t) = \gamma\sqrt{t/r} \, \exp(-\lambda^2/2) + (1-\gamma)c\lambda^2$.

The slightly modified problem

$$(2.4) \qquad \int h(\frac{|W(T)|}{(T+r)^{1/2}}, T+r) \, d\bar{Q} = \min$$

with $h(\lambda,t) = \gamma\sqrt{t/r} \, \exp(-\lambda^2/2) + (1-\gamma)c(\lambda^2 - \log t/r)$ has $S^* = \inf\{t>0 \mid |W(t)| \geq [(t+r)(\log(\frac{t+r}{r}) + 2 \log b)]^{1/2}\}$ with $b = \frac{\gamma}{2(1-\gamma)c}$ as optimal solution.

This can be seen as follows. With the help of (2.2) we get

(2.5) $\int h(\frac{|W(T)|}{(T+r)^{1/2}}, T+r)d\bar{Q} = \int \ell(\frac{d\bar{Q}_T}{dP_{o,T}})d\bar{Q}$

with $\ell(\lambda) = \gamma\lambda^{-1} + 2(1-\gamma)c \log \lambda$. The function $\ell(\lambda)$ for $\lambda>0$ has a unique minimum at $\lambda(c)=\gamma/(2(1-\gamma)c)$. Let $S^*=\inf\{t>0 \mid \frac{d\bar{Q}_t}{dP_{o,t}} \geq \lambda(c)\}$. Since $\bar{Q}\{S^*<\infty\}=1$ we get from (2.5) that S^* minimizes the risk (2.4). As we know already from Section 1, S^* is defined by the nearly parabolic boundary (1.4).

The optimal stopping rule for the risk (2.1) is bounded by S^* since for all t>s and x,y∈IR

$h(y,t)-h(x,s) \leq g(y,t)-g(x,s)$ holds.

This yields one half of the proof of Theorem 1.1. The other half is more complex and will be given in the overnext section. There the method is less ad-hoc and the statistical background of the problem is studied thoroughly.

The optimal stopping problem

(2.6) $\int g(\frac{|W(T)|}{\sqrt{T+r}}, T+r)d\bar{Q} = \min$

is closely connected with a free boundary problem for the backward diffusion equation. The minimal posterior loss at the space-time point (x,t) is given by $u(x,t)=\inf E_{\bar{Q}}^{(x,t)} g(\frac{|W(T)|}{\sqrt{T+r}}, T+r)$ where the infimum is taken over all stopping times T of the process $(W(v),v)$ which starts at (x,t). Since the infimum includes also the constant stopping time $T_t \equiv t$, it follows that $u \leq \tilde{g}$, where $\tilde{g}(x,t)=g(\frac{|x|}{\sqrt{t+r}}, t+r)$. By using similar arguments as Chernoff (1972) we show that u satisfies the equations

(2.7) $\partial_t u + \frac{1}{2}\partial_x^2 u + (1-\gamma(x,t))\frac{x}{t+r} \partial_x u = 0$

on the set $\Gamma=\{(x,t) \mid u(x,t)<\tilde{g}(x,t)\}$, with $\gamma(x,t)$ defined in the formula (1.7),

$u = \tilde{g}$ on Γ^c,

$\partial_x u = \partial_x \tilde{g}$ on the boundary $\partial \Gamma$.

These equations establish a free boundary problem corresponding to the original optimal stopping problem.

In Section 4 precise bounds for the free boundary are derived. If we consider instead of the problem (2.7) the related one where g is substituted by h, the free boundary turns out to be exactly

$$(2.8) \quad \psi(t) = \pm \left[(t+r) \left(\log \left(\frac{t+r}{r} \right) + 2 \log b \right) \right]^{1/2} .$$

3. Construction of tests of power one from smooth priors and the law of the iterated logarithm for posterior distributions

The simple Bayes rules defined in Section 1 are constructed from a prior which has a point mass at $\theta=0$ and a smooth normal part on $\{\theta\neq0\}$. Here it is pointed out that it is also possible to construct a stopping time with the properties (1.1) and (1.2) from a smooth prior. This can be done by stopping when the posterior mass of a neighbourhood of $\theta=0$ becomes too small.

Let $G_{x,t}$ denote the posterior distribution with respect to the prior $\phi(\sqrt{r}\theta)\sqrt{r}d\theta$ given that the process $(W(v),v)$ has reached (x,t). $G_{x,t}=N(\frac{x}{t+r},\frac{1}{t+r})$ where $N(\mu,\sigma^2)$ denotes the normal distribution with mean μ and variance σ^2. Let $\Phi(x)=\int_{-\infty}^{x}\phi(y)dy$. Let $\psi(t)$ be a function with the properties 1) $\psi(t)>c_o>0$ for all t, 2) $\psi(t)$ is an upper class function of the Brownian motion at infinity and 3) $\psi(t)=o(t)$. Let $J(t)=\{\theta\,|\,|\theta|\leq\frac{\psi(t)}{t+r}\}$. Then the stopping time

$$T=\inf\{t>0\,|\,G_{W(t),t}(J(t))\leq\frac{1}{2}-\Phi(-2\frac{\psi(t)}{\sqrt{t+r}})\}$$

has the properties (1.1) and (1.2). This can be seen easily by noting that T can also be written as $T=\inf\{t>0\,|\,|W(t)|\geq\psi(t)\}$.

This equivalence suggests that also a law of the iterated logarithm for posterior distributions holds. In fact this is true and it can be stated as follows. Let $I_\theta(a)=(a-\theta,a+\theta)$ denote the interval of length 2θ with midpoint a. Then the law of iterated logarithm for posterior distributions states that

$$\liminf_{t\to\infty} G_{W(t),t}\left[I_{\theta_o}(d\sqrt{\frac{\log\log t}{t}})\right]=\begin{cases} 1 \\ 0 \end{cases} \text{ if } \begin{array}{l} d>\sqrt{2} \\ d\leq\sqrt{2} \end{array}$$

P_{θ_o} - almost surely.

For more information on this type of theorem see Lerche (1981), (1982). In the next section where the optimal tests for the risk (1.1) are studied in detail, it will turn out that the construction described here plays no role at all. The simple Bayes rules which are constructed from priors with a point mass at $\theta=0$ are the important stopping times there.

4. Exact results about the shape

We consider the problem stated in (1.1): for every $0<\gamma<1$ and $c>0$ find a stopping rule T_c^* which minimizes the risk

(4.1) $\quad \rho(T) = \gamma P_0(T<\infty) + (1-\gamma)c \int_{-\infty}^{\infty} \theta^2 E_\theta T \phi(\sqrt{r}\theta)\sqrt{r}d\theta.$

At first we state that an optimal (Bayes) stopping rule T_c^* exists. Then we derive its shape. Theorem 4.2 gives upper and lower bounds for T_c^* which yields the asymptotic shape when $c\to0$ or $t\to\infty$. Theorem 4.3 refines these bounds such that a $o(c)$-approximation of the minimal Bayes-risk can be derived in Theorem 4.4. Theorem 4.5 discusses the one-sided case.

We need the following notations. The Brownian motion W with drift θ starting at time t in point x is understood as a measure $P_\theta^{(x,t)}$ on the space $C[t,\infty)$ of continuous functions on $[t,\infty)$. F_s^t denotes the σ-algebra on $C[t,\infty)$ which is generated by $W(u)$, $t\le u\le s$. The restriction of the measure $P_\theta^{(x,t)}$ on F_s^t is denoted by $P_{\theta,s}^{(x,t)}$. This notation is also used for stopping times S instead of fixed times s. When the process starts at 0 at time 0, then we very often skip the superindex and write just F_s, $P_{\theta,s}$ etc.. The Borel σ-algebra on the parameter space $\mathbb{R}$ is denoted by B. For $F=\gamma\delta_0+(1-\gamma)\int\sqrt{r}\phi(\sqrt{r}\theta)d\theta$ let $dP=dP_\theta F(d\theta)$ and $d\bar{P}=d(\int P_\theta F(d\theta))$ be its projection. Let $F_{x,t}$ denote the posterior distribution given that the process $W(t)=x$. This means that for $A\times B\in F_t\otimes B$ $\int_A F_{W(t),t}(B)\bar{P}(dW)=P(A\times B)$ holds. Thus the Bayes risk (1.1) can be rewritten as

(4.2) $\quad \rho(T) = \int_{\{T<\infty\}}(F_{W(T),T}(\{0\})+cT\int_{-\infty}^{\infty}\theta^2 F_{W(T),T}(d\theta))d\bar{P}.$

Let $\bar{P}^{(x,t)}$ denote the conditional distribution of $\bar{P}$ given that the process $W(t)=x$. It can be represented as $\bar{P}^{(x,t)}=\int P_\theta^{(x,t)}F_{x,t}(d\theta)$.

We define the posterior risk at the space-time point (x,t) for a stopping rule $T\ge t$ as

(4.3) $\quad \rho(x,t,T) = \int_{\{T<\infty\}}(F_{W(T),T}(\{0\})+c(T-t)\int_{-\infty}^{\infty}\theta^2 F_{W(T),T}(d\theta))d\bar{P}^{(x,t)}$

The minimal posterior risk at (x,t) is defined as

$$(4.4) \qquad \rho(x,t) = \inf_{T} \rho(x,t,T),$$

where the infimum is taken over all stopping times of the process $(W(s),s)$ starting at (x,t), including $T_t \equiv t$. For T_t the risk is given by

$$(4.5) \qquad \gamma(x,t) = \rho(x,t,T_t) = F_{x,t}(\{0\})$$

and therefore the inequality $\rho(x,t) \leq \gamma(x,t)$ holds. The quantity $\rho(x,t,T) + ct\int\theta^2 F_{x,t}(d\theta)$ represents the loss when the process runs without stopping up to (x,t) and then is stopped at $T \geq t$.

The following theorem states that an optimal (Bayes) stopping rule exists which minimizes (4.3) and characterizes it.

Let $C^*(c) = \{(y,s) \mid \rho(y,s) < \gamma(y,s)\}$ and

$$(4.6) \qquad T_c^* = \inf\{s \mid (W(s),s) \notin C^*(c)\}.$$

Theorem 4.1: _The stopping rule_ T_c^* _(≥t) of the space-time process_ $(W(t),t)$ _minimizes the risk (4.3) for all starting points_ (x,t).

This type of result is well known. Its statement is usually called the principle of dynamic programming. The result follows from the theory of optimal stopping for Markov processes (cf. Shiryayev (1978), p. 127) applied to the space-time process $(W(t),t)$. We note that $W(t)$ under the measure $\bar{P}$ is a diffusion process which satisfies the stochastic differential equation $dW(t) = (1-\gamma(W(t),t))\frac{W(t)}{t+r} dt + dX(t)$ where $X(t)$ is a standard Brownian motion (cf. Liptser-Shiryayev (1977), p. 258).

The stopping risk by (1.7), (1.8) and (4.5) is given by

$$(4.7) \qquad \gamma(x,t) = \frac{\gamma}{\gamma + (1-\gamma)g(x,t)} \qquad \text{with}$$

(4.8) $g(x,t) = \sqrt{\frac{r}{t+r}} \exp(\frac{x^2}{2(t+r)})$.

We note that on $\{\theta \neq 0\}$ with $G_{x,t} = N(\frac{x}{t+r}, \frac{1}{t+r})$

(4.9) $F_{x,t}(d\theta) = (1-\gamma(x,t))G_{x,t}(d\theta)$ holds.

The exact calculation of the minimal posterior risk $\rho(x,t)$ seems to be impossible for this problem. We can only derive upper and lower bounds for it. To get those we will rewrite the posterior risk in an appropriate form.

Lemma 4.1:

(4.10) $\rho(x,t,T) = \gamma(x,t)P_o^{(x,t)}$ $(t \leq T < \infty) +$

$+ (1-\gamma(x,t))c\int\theta^2 E_\theta^{(x,t)} (T-t)G_{x,t}(d\theta)$.

Remark: The posterior risk has the same form as the Bayes risk (4.1), with the slight difference that the process starts in the space-time point (x,t), stops at $T \geq t$ and has as prior $F_{x,t} = \gamma(x,t)\delta_o + (1-\gamma(x,t))G_{x,t}$, the posterior at the point (x,t).

Proof: Let $S \geq t$ be an arbitrary stopping time of Brownian motion starting at (x,t). Let F_S^t denote the σ-algebra on $C[t,\infty)$ which is generated by the process $W(u)$, $t \leq u \leq S$. Remember now the notation $\overline{P}^{(x,t)} = \int P_\theta^{(x,t)} F_{x,t}(d\theta)$. The lemma will follow from (4.3) and the subsequent statements:

(4.11) $F_{W(S),S}(\{0\})d\overline{P}^{(x,t)} = \gamma(x,t)dP_o^{(x,t)}$ holds on F_S^t, and

(4.12) $F_{W(S),S}(d\theta)d\overline{P}^{(x,t)} = dP_\theta^{(x,t)}F_{x,t}(d\theta)$

$= (1-\gamma(x,t))dP_\theta^{(x,t)}G_{x,t}(d\theta)$

holds on the σ-algebra $F_S^t \otimes B\cap\{\theta \neq 0\}$.

These are consequences of the following basic fact about posterior distributions.

The posterior of Brownian motion starting at (x,t) with prior $F_{x,t}$ at point $(W(S),S)$ is given by $F_{W(S),S}$. This is stated by the following equation:

$$(4.13) \quad F_{W(S),S}(d\theta) = \frac{dP_{\theta,S}^{(o,o)}}{dP_{o,S}^{(o,o)}} F(d\theta) \Bigg/ \int_{\textcircled{H}} \frac{dP_{\theta,S}^{(o,o)}}{dP_{o,S}^{(o,o)}} F(d\theta)$$

$$= \frac{dP_{\theta,S}^{(x,t)}}{dP_{o,S}^{(x,t)}} F_{x,t}(d\theta) \Bigg/ \int_{\textcircled{H}} \frac{dP_{\theta,S}^{(x,t)}}{dP_{o,S}^{(x,t)}} F_{x,t}(d\theta) \quad .$$

(4.13) now yields for $A \times B \in F_S^t \otimes B \cap \{\theta \neq 0\}$

$$\int_A F_{W(S),S}(B) \, d\overline{P}^{(x,t)} = \int_B P_{\theta,S}^{(x,t)}(A) F_{x,t}(d\theta)$$

which proves (4.12). In the same way (4.11) follows. □□□

The continuation region $C^*(c)$ of the optimal stopping rule for the Bayes-risk (1.1) is now approximated by upper and lower bounding regions of the space-time plane. These bounds are refined in Theorem 4.3. The bounding regions are given by sets of the type $C(\lambda) := \{(x,t) \mid \gamma(x,t) > \lambda\}$.

Theorem 4.2: There exists a constant $M > 2$ such that for every $c > 0$

$$(4.14) \quad C\left(\frac{Mc}{1+Mc}\right) \subset C^*(c) \subset C\left(\frac{2c}{1+2c}\right) \quad \textit{holds.}$$

Remark 1: Let $T_\lambda = \inf\{t > 0 \mid (W(t),t) \notin C(\lambda)\}$ then (4.14) translates to

$$T_{\frac{Mc}{1+Mc}} \leq T_c^* \leq T_{\frac{2c}{1+2c}}$$

Remark 2: The theorem holds also for the more general prior $\sqrt{r}\phi(\sqrt{r}(\theta-\mu)d\theta$ by exactly the same arguments.

Proof: At first we prove the lower inclusion of (4.14), which is the more difficult part. We show that for all points $(x,t)\in C(\frac{Mc}{1+Mc})$ (M will be specified during the proof) there exist stopping times $S_{(x,t)}$ of the process $(W(s),s)$ starting at (x,t), such that

(4.15) $\rho(x,t,S_{(x,t)}) < \gamma(x,t)$

holds.

Since by definition $\rho(x,t)\leq\rho(x,t,S_{(x,t)})$, it follows from (4.15) and Theorem 1 that $(x,t)\in C^*(c)$.

We choose the stopping times as

$$S_{(x,t)} = \inf\{s>t \mid \gamma(W(s),s)\leq Qc\}$$

where the constant Q>1 will be defined below. We assume also that Qc<1. (In fact the stopping times $S_{(x,t)}$ all arise from the same stopping time T_{Qc} by changing the starting point of the process.) We need several representations of $S_{(x,t)}$ during the proof.

(4.16) $S_{(x,t)} = \inf\{s>t \mid F_{W(s),s}(\{0\}) \leq Qc\}$

$$= \inf\{s>t \mid \int \frac{dP_{\theta,s}^{(x,t)}}{dP_{0,s}^{(x,t)}} G_{x,t}(d\theta) \geq b(x,t)\}$$

$$= \inf\{s>t \mid \sqrt{\frac{t+r}{s+r}} \exp(\frac{1}{2}(\frac{W(s)^2}{s+r} - \frac{x^2}{t+r})) \geq b(x,t)\}$$

where $b(x,t) = \frac{\gamma(x,t)(1-Qc)}{(1-\gamma(x,t))Qc}$.

The first equality holds by definition, the second equality follows by calculation since by (4.13)

$$F_{W(s),s}(\{0\}) = \frac{\gamma(x,t)}{\gamma(x,t)+(1-\gamma(x,t))h(x,t,w,s)}$$

with $\quad h(x,t,w,s) = \int \frac{dP_{\theta,s}^{(x,t)}}{dP_{o,s}^{(x,t)}} G_{x,t}(d\theta) \quad .$

The third equality follows by the subsequent calculation (note here that $G_{x,t}=N(\frac{x}{t+r},\frac{1}{t+r}))$.

$$\int \frac{dP_{\theta,s}^{(x,t)}}{dP_{o,s}^{(x,t)}} G_{x,t}(d\theta) =$$

$$= \int e^{\theta(W(s)-x)-\frac{1}{2}\theta^2(s-t)} \sqrt{\frac{t+r}{2\pi}} \, e^{-\frac{(t+r)}{2}(\theta-\frac{x}{t+r})^2} \, d\theta$$

$$= \sqrt{(t+r)} \, e^{-\frac{x^2}{2(t+r)}} \int e^{\theta W(s)-\frac{1}{2}\theta^2(s+r)} \frac{d\theta}{\sqrt{2\pi}}$$

$$= \sqrt{\frac{t+r}{s+r}} \, \exp\left(\frac{1}{2}\left(\frac{W(s)^2}{s+r} - \frac{x^2}{t+r}\right)\right).$$

We start now to estimate the posterior risk for $S_{(x,t)}$, which is given according to Lemma 1 by

$$(4.17) \quad \rho(x,t,S_{(x,t)}) = \gamma(x,t)P_o^{(x,t)}(t<S_{(x,t)}<\infty)$$

$$+ (1-\gamma(x,t))c\int\theta^2(S_{x,t}-t)dQ^{(x,t)}$$

with the new notation $Q^{(x,t)}(dW,d\theta)=P_\theta^{(x,t)}(dW)G_{x,t}(d\theta)$. We will also use $\bar{Q}^{(x,t)}(dW)=\int P_\theta^{(x,t)}(dW)G_{x,t}(d\theta)$ and will from now on simply write S instead of $S_{(x,t)}$. Then we get for the first term

$$(4.18) \quad P_o^{(x,t)}(t<S<\infty) = \frac{Qc}{1-Qc} \frac{1-\gamma(x,t)}{\gamma(x,t)} = b(x,t)^{-1} \quad .$$

This follows from a well known martingale argument (see e.g. Theorem 2,1, Chap. I) by using (4.16):

$$P_o^{(x,t)}(t<S<s_o) = \int_{\{t<S<s_o\}} \frac{dP_{o,S}^{(x,t)}}{d\bar{Q}_S^{(x,t)}} \, d\bar{Q}^{(x,t)}$$

$$= b(x,t)^{-1} \bar{Q}^{(x,t)}\{t<S<s_o\}.$$

Since $\bar{Q}^{(x,t)}\{t<S<s_o\} \to 1$ as $s_o \to \infty$, (4.18) follows. We note that for $(x,t) \in C(Qc)$, $b(x,t)>1$ holds and that thus the probability in (4.18) is less than one.

To estimate the second term of (4.17) we rewrite the integral. Since on F_S^t we have

$$Q^{(x,t)}(dW,d\theta) = N\left(\frac{W(S)}{S+r}, \frac{1}{S+r}\right)(d\theta)\bar{Q}^{(x,t)}(dW),$$

we get for the integral

$$(4.19) \quad \int \theta^2 (S-t) \, dQ^{(x,t)} = \int \theta^2 (S+r) \, dQ^{(x,t)} - \int \theta^2 (t+r) \, dQ^{(x,t)}$$

$$= \int\int \theta^2 (S+r) N\left(\frac{W(S)}{S+r}, \frac{1}{S+r}\right)(d\theta) \, d\bar{Q}^{(x,t)}$$

$$- \int \theta^2 (t+r) G_{x,t}(d\theta)$$

$$= \int \left(\frac{W(S)^2}{S+r} - \frac{x^2}{t+r}\right) d\bar{Q}^{(x,t)}$$

Using now the third form in (4.16) of the stopping rule S yields

$$(4.20) \quad \int \left(\frac{W(S)^2}{S+r} - \frac{x^2}{t+r}\right) d\bar{Q}^{(x,t)} = 2\log b(x,t) + \int \log\left(\frac{S+r}{t+r}\right) d\bar{Q}^{(x,t)}$$

Let $\alpha>3$. We show now that there exists a constant $0<C_\alpha<\infty$ with

$$(4.21) \quad \int \log\left(\frac{S+r}{t+r}\right) d\bar{Q}^{(x,t)} \leq C_\alpha \left[\int \theta^2 (S-t) \, dQ^{(x,t)}\right]^{1/\alpha}$$

Then (4.19), (4.20) and (4.21) yield

$$(4.22) \quad \int \theta^2 (S-t) dQ^{(x,t)} \leq 2\log b(x,t) + C_\alpha \left[\int \theta^2 (S-t) dQ^{(x,t)} \right]^{1/\alpha}$$

from which one can derive (4.15) as will be explained below.

The proof of (4.21) runs as follows. By using the inequality $\log(1+x) \leq K_\alpha x^{1/\alpha}$ for $x \geq 0$ we get by Hölder's inequality

$$(4.23) \quad \int \log\left(\frac{S+r}{t+r}\right) d\bar{Q}^{(x,t)} = \int \log\left(1+\frac{S-t}{t+r}\right) d\bar{Q}^{(x,t)}$$

$$\leq K_\alpha \int \left(\frac{S-t}{t+r}\right)^{1/\alpha} d\bar{Q}^{(x,t)}$$

$$= K_\alpha \int (\theta^2 (S-t))^{1/\alpha} (\theta^2 (t+r))^{-1/\alpha} d\bar{Q}^{(x,t)}$$

$$\leq K_\alpha (\int \theta^2 (S-t) dQ^{(x,t)})^{1/\alpha} (\int (\theta^2 (t+r))^{\frac{-1}{\alpha-1}} G_{x,t}(d\theta))^{\frac{\alpha-1}{\alpha}}.$$

But since $G_{x,t} = N(\frac{x}{t+r}, \frac{1}{t+r})$ we get for $\alpha > 3$

$$\int (\theta^2 (t+r))^{\frac{-1}{\alpha-1}} G_{x,t}(d\theta) \leq \int (\theta^2 (t+r))^{\frac{-1}{\alpha-1}} N(0, \frac{1}{t+r})(d\theta)$$

$$= \int y^{-\frac{2}{\alpha-1}} N(0,1)(dy) < \infty.$$

We now put $C_\alpha = K_\alpha (\int y^{-\frac{2}{\alpha-1}} N(0,1)(dy))^{\frac{\alpha-1}{\alpha}}$ and get finally (4.21) and (4.22).

Let $b>1$ be given, then by (4.22) there exists a constant $B>2$ such that for all $b(x,t) \geq b$

$$(4.24) \quad \int \theta^2 (S-t) dQ^{(x,t)} \leq B \log b(x,t) \qquad \text{holds.}$$

Now we choose $Q = \frac{B}{1+Bc}$ and $M = bB$. Then for $(x,t) \in C(\frac{Mc}{1+Mc})$ we have

$$b(x,t) = \frac{1-Qc}{Qc} \frac{\gamma(x,t)}{1-\gamma(x,t)} = \frac{\gamma(x,t)}{Bc(1-\gamma(x,t))} > \frac{Mc}{Bc} = b,$$

and by (4.17), (4.18) and (4.24) we get further

$$\rho(x,t,S) \leq \gamma(x,t)b(x,t)^{-1} + (1-\gamma(x,t))Bc \log b(x,t)$$

$$= \gamma(x,t)b(x,t)^{-1}(1+\log b(x,t))$$

$$< \gamma(x,t) .$$

The last inequality follows from the inequality $x(1+\log x^{-1})<1$ for $x<1$ since $b(x,t)>b>1$. This proves (4.15).

Now we prove the upper inclusion. We show

$$(4.25) \quad \rho(x,t) = \gamma(x,t) \quad \text{if} \quad \gamma(x,t) \leq \frac{2c}{1+2c} .$$

Statement (4.25) implies the upper inclusion of statement (4.14). The method of proof of (4.25) consists of comparing the Bayes rule T_c^* with the best rule if θ would be known.

For the Bayes rule T_c^* we always have

$$\gamma(x,t) \geq \rho(x,t) = \rho(x,t,T_c^*)$$

$$= \gamma(x,t)P_o^{(x,t)}(t \leq T_c^* < \infty) + (1-\gamma(x,t)).c\int \theta^2 (T_c^*-t)\,dQ^{(x,t)}$$

$$= \int_{-\infty}^{\infty}\left[\gamma(x,t)P_o^{(x,t)}(t \leq T_c^* < \infty) + (1-\gamma(x,t))c\theta^2 E_\theta^{(x,t)}(T_c^*-t)\right]G_{x,t}(d\theta) .$$

Let the process W start in x and let W(u)=z. Under the transformation $y=\theta(z-x)$, $s=\theta^2(u-t)$ Brownian motion with drift θ (resp. 0) goes over into Brownian motion with drift 1 (resp. 0). With $S_\theta=\theta^2(T_c^*-t)$ we get

$$= \int_{-\infty}^{\infty}\left[\gamma(x,t)P_o^{(o,o)}(0 \leq S_\theta < \infty) + (1-\gamma(x,t))cE_1^{(o,o)}S_\theta\right]G_{x,t}(d\theta)$$

$$> \inf_S\left[\gamma(x,t)P_o^{(o,o)}(0 \leq S < \infty) + (1-\gamma(x,t))cE_1^{(o,o)}S\right] =: \tilde{\rho}(x,t) .$$

But $\tilde{\rho}(x,t)$ is the minimal Bayes-risk of (1.5) with $\gamma=\gamma(x,t)$. We determine now its Bayes-stopping set. By (1.6)

$$\tilde{\rho}(x,t) = \min_{0 \le p \le 1} (\gamma p + 2(1-\gamma)c \log p^{-1})$$

$$= \gamma p_0 + 2(1-\gamma)c \log p_0^{-1}$$

with $p_0 = \dfrac{2(1-\gamma)c}{\gamma} \wedge 1$ (p_0 denotes the stopping probability). Thus

$$\tilde{\rho}(x,t) = 2(1-\gamma)c\left[1+\log(\frac{2(1-\gamma)c}{\gamma})^{-1}\right] \qquad \text{if } \frac{2(1-\gamma)c}{\gamma} \le 1$$

and $\qquad \tilde{\rho}(x,t) = \gamma \qquad$ if $\dfrac{2(1-\gamma)c}{\gamma} \ge 1.$

The Bayes stopping region is therefore equal to

$$\{(x,t) \mid \gamma(x,t) = \tilde{\rho}(x,t)\} = \{(x,t) \mid \gamma(x,t) \le \gamma_0\},$$

where γ_0 is determined by the equation

$$p_0 = \frac{2(1-\gamma_0)c}{\gamma_0} = 1. \quad \text{Thus } \gamma_0 = \frac{2c}{1+2c} \quad .$$

□□□

We now derive a refinement of the statement of Theorem 4.2. For this we need a somewhat more general notation. If $h(t)$ is a positive function of time we shall denote by $C(h(.))=\{(x,t) \mid \gamma(x,t)>h(t)\}$.

Theorem 4.3. For every $c>0$ there exists a bounded function $\tilde{c}(.) \ge c$ with

(a) $\tilde{c}(t)/c \to 1$ when $t \to \infty$ for every fixed c, and
(b) $\sup_{0<t<\infty} \tilde{c}(t)/c \to 1$ when $c \to 0$,

such that

(4.26) $\quad C(\dfrac{2\tilde{c}(.)}{1+2\tilde{c}(.)}) \subset C^{\star}(c) \subset C(\dfrac{2c}{1+2c}) \qquad$ *holds.*

The theorem states that for c small or t large the optimal stopping region is very near to its upper bound $C(\frac{2c}{1+2c})$. The proof of Theorem 4.3 will show that the upper bound of $\check{c}(.)/c$ is a bit larger than M/2 where M is the constant appearing in Theorem 4.2.

Several conclusions can be drawn from this theorem. Let $\psi_c^*(t) = \inf\{x>0 \mid \rho(x,t)=\gamma(x,t)\}$. By Theorem 4.2 this definition makes sense. Thus by the symmetry of the problem

$$T_c^* = \inf\{t \mid |W(t)| \geq \psi_c^*(t)\} \quad .$$

Corollary 4.1:

$$\psi_c^*(t) = \left[(t+r)(\log(\frac{t+r}{r}) + 2 \log \frac{\gamma}{2(1-\gamma)c} + o(1)) \right]^{1/2}$$

when $t \to \infty$.

Corollary 4.2:

$$\psi_c^*(t) = \left[(t+r)(2 \log \frac{\gamma}{2(1-\gamma)c} + \log(\frac{t+r}{r}) + o(1)) \right]^{1/2}$$

uniformly in t when $c \to 0$.

Corollary 4.3: For every $\varepsilon>0$ there exists a $c_o>0$ such that

$$T_{\frac{2c(1+\varepsilon)}{1+2c(1+\varepsilon)}} \leq T_c^* \leq T_{\frac{2c}{1+2c}} \quad \text{for all } 0<c\leq c_o.$$

We can combine Corollary 4.3 with a result about boundary crossing distributions of Chapter I to get the minimal Bayes risk for (4.1) up to an o(c)-term. A related O(c)-result for the Bayes risk has been obtained by Pollak (1978), when there is an indifference zone in the parameter space.

Theorem 4.4:

$$(4.27) \quad 0 \leq \rho(T_{\frac{2c}{1+2c}}) - \rho(T_c^*) = o(c) \quad \text{when } c \to 0.$$

The minimal Bayes risk for (4.1) *is given by*

$$(4.28) \quad \rho(T_c^*) = 2(1-\gamma)c\left[\log b + \frac{1}{2}\log\log b + 1 + \frac{1}{2}\log 2 - A + o(1)\right]$$

when $c \to 0$. *Here* $b = \dfrac{\gamma}{2(1-\gamma)c}$ *and* $A = 2\int_0^\infty \log x\ \phi(x)dx$.

Remark: Comparing statement (4.28) with the related formula (19) for the simple testing problem, shows that the additional term $2(1-\gamma)c[\frac{1}{2}\log(2\log b) - A + o(1)]$ appears in the minimal Bayes risk. This is the price for the ignorance about the parameter $\theta \neq 0$.

Proof: From Corollary 4.3 follows

$$(4.29) \quad \rho(T_{\frac{2c(1+\varepsilon)}{1+2c(1+\varepsilon)}}) - \gamma\left[P_0(T_{\frac{2c(1+\varepsilon)}{1+2c(1+\varepsilon)}} < \infty) - P_0(T_{\frac{2c}{1+2c}} < \infty)\right]$$

$$\leq \rho(T_c^*) \leq \rho(T_{\frac{2c}{1+2c}}) \ .$$

We now show that the right and the left hand side of (4.29) differ from each other only by a $o(c)$-term. Formula (4.18) yields

$$(4.30) \quad P_0(T_{\frac{2c(1+\varepsilon)}{1+2c(1+\varepsilon)}} < \infty) - P_0(T_{\frac{2c}{1+2c}} < \infty) \leq \varepsilon b^{-1} = O(\varepsilon c).$$

Now we compute $\rho(T_{\frac{2c}{1+2c}})$. We write from now on for simplicity T instead of $T_{\frac{2c}{1+2c}}$. By (4.18) and (4.20) for $x=0$, $t=0$

$$(4.31) \quad \rho(T) = 2(1-\gamma)c\left[1+\log b + \frac{1}{2}\int\log\left(\frac{T+r}{r}\right)dQ\right]$$

The integral on the right hand side can be calculated by using Theorem 4.3 of Chapter I. The following result is intuitively plausible by virtue of the relation

$$P_\theta\{T/\log b \to 2\theta^{-2}\} = 1.$$

(4.32) $\int \log(\frac{T+r}{r})dQ = \log(2 \log b) - 2A + o(1).$

Combining (4.32) with (4.31) yields

(4.33) $\rho(T_{\frac{2c}{1+2c}}) = 2(1-\gamma)c\left[\log b + \frac{\log(2 \log b)}{2} + 1 - A + o(1)\right].$

From (4.30) and (4.33) it follows also that

(4.34) $\rho(T_{\frac{2c(1+\epsilon)}{1+2c(1+\epsilon)}}) = \rho(T_{\frac{2c}{1+2c}}) + O(\epsilon c).$

Statement (4.34) together with (4.29) and (4.30) yield (4.27) and (4.27) together with (4.33) yield (4.28).

□□□

.

Proof of Theorem 4.3: The upper inclusion of (4.26) is already proved by (4.25). Now we prove the lower one. For the stopping times

(4.35) $S_{(x,t)} = \inf\{s>t \,|\, \gamma(W(s),s) < \frac{2c}{1+2c}\}$

we show that

(4.36) $\rho(x,t,S_{(x,t)}) < \gamma(x,t)$ for $(x,t) \in C\left(\frac{2\tilde{c}(\cdot)}{1+2\tilde{c}(\cdot)}\right) \backslash C\left(\frac{Mc}{1+Mc}\right)$

where $\tilde{c}(\cdot)$ will be specified below. M is the constant of Theorem 4.2.

Then (4.14) together with (4.36) imply the lower inclusion of (4.26).

Now we define $\tilde{c}(t)$. We note that for the stopping times (4.35) by (4.22) with $\alpha=4$ and $b(x,t) = \frac{\gamma(x,t)}{2(1-\gamma(x,t))c}$ the inequality

$$(4.37) \quad \int \theta^2 (S-t) dQ^{(x,t)} < 2\log b(x,t) + C(x,t) \left[\int \theta^2 (S-t) dQ^{(x,t)} \right]^{1/4}$$

holds. The constants $C(x,t)$ are given by

$$(4.38) \quad C(x,t) = K(\int (\theta^2 (t+r))^{-1/3} G_{x,t}(d\theta))^{3/4}$$

$$= K(\int y^{-2/3} N(\hat{\theta}\sqrt{t+r},1)(dy))^{3/4}$$

with $\hat{\theta} = \frac{x}{t+r}$.

Let ψ_c^+ and ψ_c^- denote the positive and negative branches of the solution of the implicit equation $\gamma(\psi_c^{\pm}(t),t) = \frac{Mc}{1+Mc}$. By symmetry $\psi_c^{\pm} = \pm\psi_c$ where ψ_c is given by

$$\psi_c(t) = \left[(t+r)(\log(\frac{t+r}{r}) + 2\log \frac{\gamma}{(1-\gamma)Mc}) \right]^{1/2} .$$

We choose

$$(4.39) \quad e(t,c) = -\log(1-C(\psi_c(t),t)^{1/4}) \wedge \log M/2$$

and put $\bar{c}(t) = c \exp(e(t,c))$. Let $a > 1$. Let $d(a)=\inf\{y>1 \mid a \log(ay)<ay-1\}$. $d(a)$ is uniquely determined. We define $\tilde{c}(t)=d(\bar{c}(t)/c)\bar{c}(t)$.

Now we claim that $\tilde{c}(\cdot)/c$ has the demanded properties (a) and (b). By (4.38) $C(x,t)$ depends only on $|\hat{\theta}\sqrt{t+r}| = \frac{|x|}{\sqrt{t+r}}$. Evaluating $|\hat{\theta}\sqrt{t+r}|$ at the graphs $(\pm\psi_c(t),t)$ yields:

$$\left| \hat{\theta} \sqrt{t+r} \right| = \left[\log(t+r) + 2 \log \frac{\gamma}{(1-\gamma) Mc} \right]^{1/2} ,$$

which tends to infinity, uniformly in t when c → 0, or when t → ∞.
Consequently $C(\pm\psi_c(t), t)$ → 0 and therefore by (4.39) e(t,c) → 0,
uniformly in t when c → 0, or when t → ∞. Since d(a)→1 as a→1 the
properties (a) and (b) follow.

Now we show (4.36). As a first step we prove

$$(4.40) \quad c \int \theta^2 (S-t) \, dQ^{(x,t)} < 2\bar{c}(t) \log b(x,t)$$

for $(x,t) \in C(\frac{2\bar{c}(\cdot)}{1+2\bar{c}(\cdot)}) \diagdown C(\frac{Mc}{1+Mc})$.

By (4.39) we can assume that $\bar{c}(t) < Mc/2$. Let $H(x,t) = \int \theta^2 (S-t) \, d\bar{Q}^{(x,t)}$.
Then we have from (4.19), (4.20) and (4.37) (with the x and t variables
suppressed)

$$(4.41) \quad 2 \log b < H < 2 \log b + CH^{1/4}.$$

Let $C_1(t) = C(\psi_c(t), t)$. Then $0 < C(x,t) < C_1(t)$ holds on $C(\frac{Mc}{1+Mc})^c$
and therefore

$$(4.42) \quad (1 - C_1/H^{3/4}) H < 2 \log b.$$

If

$$(4.43) \quad b(x,t) > \exp(\frac{1}{2} C_1(t))$$

holds for $(x,t) \in C(\frac{2\bar{c}(\cdot)}{1+2\bar{c}(\cdot)}) \diagdown C(\frac{Mc}{1+Mc})$ then we get from the left hand
side of (4.41) $C_1(t) < H(x,t)$ and therefore from (4.42)

$$H(x,t) < 2 \log(b(x,t)) (1 - C_1(t)^{1/4})^{-1}.$$

But this yields (4.40).

It is left to show that (4.43) holds.

Let $c(t) = \dfrac{c}{1-C_1(t)^{1/4}} \,/\, \exp(\tfrac{1}{2}C_1(t))$. An elementary calculation shows

that $c(t) > c$ for $0 < C_1(t) < 1$. Then

$$b(x,t) = \frac{\gamma(x,t)}{2(1-\gamma(x,t))c} > \frac{\gamma(x,t)}{2(1-\gamma(x,t))c(t)}$$

$$= \frac{\gamma(x,t)\exp(\tfrac{1}{2}C_1(t))}{2(1-\gamma(x,t))\bar{c}(t)}$$

$$> \exp(\tfrac{1}{2}C_1(t)) .$$

The second equation holds since

$$\bar{c}(t) = c(1-C_1(t)^{1/4})^{-1} < M/2,$$

and the last inequality follows from the definition of $C(\frac{2\bar{c}(\cdot)}{1+2\bar{c}(\cdot)})$. This proves (4.43) and completes the proof of (4.40).

Combining now (4.17) and (4.18) with (4.40) yields for the stopping times (4.35) the estimate for the Bayes risk

(4.44) $\rho(x,t,S) < \gamma(x,t)b(x,t)^{-1} + 2(1-\gamma(x,t))\bar{c}(t)\log b(x,t)$

with $b(x,t) = \dfrac{\gamma(x,t)}{2(1-\gamma(x,t))c}$ on $C(\frac{2\bar{c}(\cdot)}{1+2\bar{c}(\cdot)}) \smallsetminus C(\frac{Mc}{1+Mc})$.

We assume now that $(x,t) \in C(\frac{2\tilde{c}(\cdot)}{1+2\tilde{c}(\cdot)}) \smallsetminus C(\frac{Mc}{1+Mc})$ and estimate the right hand side of (4.44) further. It is equal to

(4.45) $\gamma(x,t)b(x,t)^{-1}[1 + (\bar{c}(t)/c)\log b(x,t)]$

$= \gamma(x,t)[h(x,t)\bar{c}(t)/c]^{-1}[1+(\bar{c}(t)/c)\log(h(x,t)\bar{c}(t)/c)]$

with $h(x,t) = \gamma(x,t)/(2(1-\gamma(x,t))\bar{c}(t))$.

Since $(ay)^{-1}(1+a \cdot \log(ay))<1$ for $y>d(a)$ and since on $C(\frac{2\tilde{c}(.)}{1+2\tilde{c}(.)})$, $h(x,t)>d(\bar{c}(t)/c)$ by the definition of $\tilde{c}$, it follows that the expression of (4.45) is strictly less than $\gamma(x,t)$. This yields (4.36) and completes the proof of Theorem 4.3.

□□□

In the last part of this section we study the one-sided testing problem. We consider the Bayes risk given by

$$(4.46) \quad \rho(T) = \gamma P_o(T<\infty) + (1-\gamma)c \int_o^\infty \theta^2 E_\theta T \phi(\sqrt{r}\theta) 2\sqrt{r} \ d\theta.$$

For it we can characterize the minimizing stopping rule T_c^* (it also exists) by results similar to those for the two-sided case.

If not mentioned otherwise we will use the same notations as in the preceding part for the corresponding objects here. For instance

$$\gamma(x,t) = F_{x,t}(\{0\})$$

$$= (1+ \frac{1-\gamma}{\gamma} \int_o^\infty e^{\theta x - \frac{1}{2}\theta^2 t} \phi(\theta\sqrt{r}) 2\sqrt{r}d\theta)^{-1},$$

and also $\rho(x,t,T), \rho(x,T), C^*(c), C(Kc)$ etc.. The prior on $[0,\infty)$ is given by

$$F = \gamma\delta_o + (1-\gamma) \int \phi(\sqrt{r}\theta) 2\sqrt{r}d\theta.$$

The posterior at (x,t) can be represented as
$F_{x,t} = \gamma(x,t)\delta_o + (1-\gamma(x,t))H_{x,t}$ where $H_{x,t} = N(\frac{x}{t+r}, \frac{1}{t+r})/\Phi(\frac{x}{\sqrt{t+r}})$ on $(0,\infty)$.
We only state the analogous result to Theorem 4.2. The counterpart to Theorem 4.3 holds also and can be proved exactly in the same way as Theorem 4.3.

Theorem 4.5: There exists a constant K>2 such that for every c>0

$$(4.47) \quad C(\frac{Kc}{1+Kc}) \subset C^*(c) \subset C(\frac{2c}{1+2c}) \quad holds.$$

Proof: The proof of the upper inclusion of (4.47) runs exactly along the same lines as that of (4.14). For the lower inclusion we show that for all $(x,t) \in C(Kc)$ there exists a stopping time $S_{(x,t)}$ of the process $(W(s),s)$ starting at (x,t) such that

$$\rho(x,t,S_{(x,t)}) < \gamma(x,t) \qquad \text{holds.}$$

Let Q denote a constant which satisfies Qc<1. We choose

$$S_{(x,t)} = \inf\{s>t \mid \gamma(W(s),s) \leq Qc\}$$

which can be rewritten as

$$(4.48) \quad S_{(x,t)} = \inf\{s>t \mid \int_0^\infty \frac{dP_{\theta,s}^{(x,t)}}{dP_{0,s}^{(x,t)}} H_{x,t}(d\theta) \geq b(x,t)\}$$

$$= \inf\{s>t \mid \sqrt{\frac{t+r}{s+r}} \exp\left(\frac{1}{2}\left(\frac{W(s)^2}{s+r} - \frac{x^2}{t+r}\right)\right) \frac{\Phi\left(\frac{W(s)}{\sqrt{s+r}}\right)}{\Phi\left(\frac{x}{\sqrt{t+r}}\right)} \geq b(x,t)\}$$

with $\quad b(x,t) = \frac{\gamma(x,t)(1-Qc)}{(1-\gamma(x,t))Qc} \quad$ and $\quad \Phi(y) = \int_{-\infty}^y \phi(x)\,dx.$

The posterior risk at (x,t) can be represented as

$$(4.49) \quad \rho(x,t,S_{(x,t)}) = \gamma(x,t) P_0^{(x,t)}(t<S_{(x,t)}<\infty) +$$

$$(1-\gamma(x,t))c \int_0^\infty \theta^2 E_\theta^{(x,t)}(S_{(x,t)}-t) H_{x,t}(d\theta).$$

From here on we write S instead of $S_{(x,t)}$. The same martingale argument as for (4.18) yields

$$P_0^{(x,t)}(t<S<\infty) = b(x,t)^{-1}.$$

The estimate of the other part of the Bayes-risk (4.49) is a bit more complicated as that of the corresponding part of (4.17). It can be expressed after some calculations similar to those of (4.19) as follows:

$$(4.50) \quad \int \theta^2 E_\theta^{(x,t)}(S-t) H_{x,t}(d\theta) =$$

$$= \int (\frac{W(S)^2}{S+r} - \frac{x^2}{t+r}) d\overline{Q}^{(x,t)} + \int (h(\frac{W(S)}{\sqrt{S+r}}) - h(\frac{x}{\sqrt{t+r}})) d\overline{Q}^{(x,t)}$$

with $h(y) = y\phi(y)/\Phi(y)$ and $\overline{Q}^{(x,t)} = \int_0^\infty P_\theta^{(x,t)} H_{x,t}(d\theta)$.

Using the defining equation (4.48) of the stopping time S yields

(4.51) $\int_0^\infty \theta^2 E_\theta^{(x,t)} (S-t) H_{x,t}(d\theta) =$

$$= 2 \log b + \int \log(\frac{S+r}{t+r}) d\overline{Q}^{(x,t)}$$

$$+ \int \left[h(\frac{W(S)}{\sqrt{S+r}}) - h(\frac{x}{\sqrt{t+r}}) \right] d\overline{Q}^{(x,t)}$$

$$- 2\int \left[g(\frac{W(S)}{\sqrt{S+r}}) - g(\frac{x}{\sqrt{t+r}}) \right] d\overline{Q}^{(x,t)}$$

with $g(y) = \log \Phi(y)$.
Now after some calculations we get

(4.52) $h(\frac{W(S)}{\sqrt{S+r}}) - h(\frac{x}{\sqrt{t+r}}) - 2\left[g(\frac{W(S)}{\sqrt{S+r}}) - g(\frac{x}{\sqrt{t+r}}) \right] = \int_{\frac{x}{\sqrt{t+r}}}^{\frac{W(S)}{\sqrt{S+r}}} (h'(y) - 2g'(y)) dy$

$$= \int_{\frac{x}{\sqrt{t+r}}}^{\frac{W(S)}{\sqrt{S+r}}} \frac{\phi(y)}{\Phi(y)} \left[-(1+y^2) - h(y) \right] dy.$$

But this integral is always negative. It is obvious that the integrand is negative for positive values of y. That it is also negative for negative y-values can be seen as follows. We have to show that

(4.53) $-(1+y^2) - y\phi(y)/\Phi(y) \leq 0$ for $y < 0$

which is equivalent to

$$-(1+y^2)(1-\Phi(y)) + y\phi(y) \leq 0 \quad \text{for } y > 0$$

and to

(4.54) $1-\phi(y) \geq (y/(1+y^2))\phi(y)$ for y>0.

Both sides of (4.54) vanish at y=∞ and the derivative of the left hand side is always smaller than that of the right hand side and both are negative, i.e.

$$-\phi(y) \leq -\phi(y)(1-2/(1+y^2)^2)$$ for all y>0.

This yields (4.53) and therefore the integrand in (4.52) is always negative.

It is left to show that $\frac{W(S)}{\sqrt{S+r}} > \frac{x}{\sqrt{t+r}}$. Now let $\frac{K}{1+Kc} > Q$. Then $(x,t) \in C(\frac{Kc}{1+Kc})$ implies $\gamma(x,t) > Qc$, which yields b(x,t)>1. This together with (4.48) implies for S>t the inequality

$$e^{\frac{x^2}{2(t+r)}} \phi(\frac{x}{\sqrt{t+r}}) < \sqrt{\frac{t+r}{S+r}} \; e^{\frac{W(S)^2}{2(S+r)}} \phi(\frac{W(S)}{\sqrt{S+r}}) \; .$$

Since the function $\lambda \mapsto e^{\lambda^2/2}\phi(\lambda)$ is increasing this yields $\frac{W(S)}{\sqrt{S+r}} > \frac{x}{\sqrt{t+r}}$. Thus the expression in (4.52) is always negative which by (4.51) yields

$$\int_0^\infty \theta^2 E_\theta^{(x,t)} (S-t) H_{x,t}(d\theta) \leq 2 \log b + \int \log(\frac{S+r}{t+r}) d\bar{Q}^{(x,t)} \; .$$

The rest of the proof runs similarly to that of Theorem 4.2 from (4.20) on.

□□□

5. An optimal property of the repeated significance test

We consider the problem of testing the sign of the drift θ of a Brownian
motion. As in the preceding section we let the costs depend on the
underlying parameter and choose it as "$c\theta^2$", $c>0$. We show that a certain
simple Bayes rule, which defines a repeated significance test, is optimal for the testing problem in a Bayes sense. The simple Bayes rules
stop sampling when the posterior mass of the hypothesis or the alternative is too small.

It is well known that also Wald's sequential probability ratio test is
defined by a simple Bayes rule and that it is optimal for simple hypotheses. We point out that the repeated significance test (RST) is the
natural counterpart to Wald's sequential probability ratio test (SPRT)
for testing composite hypotheses without an indifference zone.

The parameter sets of the hypothesis H_O and the alternative H_1 are
given by $\Theta_O=\{\theta<0\}$ and by $\Theta_1=\{\theta>0\}$. We assume 0-1 loss, the usual loss
structure for testing. The observation costs are choosen as $c\theta^2$ where
c is a positive constant and θ is the drift of the observed Brownian
motion. On the parameter space $\Theta_O \cup \Theta_1$ we put the normal prior
$G(d\theta)=\phi(\sqrt{r}(\theta-\mu))\cdot\sqrt{r}d\theta$ with $\phi(y) = \frac{1}{\sqrt{2\pi}} e^{-y^2/2}$. The Bayes risk for a decision procedure (T,δ) consisting of a stopping rule T of the Brownian
motion and a terminal decision rule δ, is given by

$$(5.1) \quad \rho(T,\delta) = \int_{-\infty}^{0}(P_\theta\{H_O \text{ rejected } (\delta)\}+c\theta^2 E_\theta T)G(d\theta)$$

$$+ \int_{0}^{\infty}(P_\theta\{H_1 \text{ rejected } (\delta)\}+c\theta^2 E_\theta T)G(d\theta).$$

Let $G_{x,t}$ denote the posterior distribution of θ, given that the process
$(W(s),s)$ has reached (x,t). It is equal to $G_{x,t}=N(\frac{x+r\mu}{t+r} , \frac{1}{t+r})$. Here
$N(\rho,\sigma^2)$ denotes the normal distribution with mean ρ and variance σ^2.
For $\lambda>0$ the simple Bayes rule is defined as

$$T_\lambda = \inf\{t>0 \mid \min_{i=0,1} G_{W(t),t}(\Theta_i) \le \Phi(-\lambda)\},$$

where Φ denotes the standard normal distribution function. It can also be expressed as

$$T_\lambda = \inf\{t>0 \mid \frac{|W(t)+r\mu|}{\sqrt{t+r}} \geq \lambda\}.$$

The following result states that a simple Bayes rule is optimal for the risk (5.1). The corresponding stopping boundary defines a repeated significance test ($\mu=0$ is the usual case).

Theorem 5.1: Let $0<c<\infty$. Let $\lambda(c)$ denote the solution of the equation $\phi(\lambda)/\lambda=2c$ and let $T^*=T_{\lambda(c)}$. Let δ^* denote the final decision rule which rejects the hypothesis if and only if $W(T^*)>-r\mu$. Let $|\mu|\sqrt{r}\leq\lambda(c)$. Then the procedure (T^*,δ^*) minimizes the Bayes risk (5.1).

Proof: Let $\bar{Q}=\int_{-\infty}^{\infty}P_\theta\phi(\sqrt{r}(\theta-\mu))\sqrt{r}d\theta$. Then $P_{\theta,t}(dW)\phi(\sqrt{r}(\theta-\mu))\sqrt{r}d\theta=G_{W(t),t}(d\theta)\bar{Q}(dW)$. A well-known argument yields that $\rho(T,\delta^*)\leq\rho(T,\delta)$ for every stopping time T. Let $r(T,\delta)$ denote the part of the Bayes risk (5.1) consisting of the error probabilities. Then

(5.2) $$r(T,\delta^*)=\int \min_{i=0,1} G_{W(T),T}(\Theta_i)d\bar{Q}=\int\phi(-\frac{|W(T)+r\mu|}{\sqrt{T+r}})d\bar{Q} .$$

On the other hand, the Bayes formula and Fubini's theorem yield

(5.3) $$\int_{-\infty}^{\infty}\theta^2 E_\theta T\phi(\sqrt{r}(\theta-\mu))\sqrt{r}d\theta$$

$$=\int_{-\infty}^{\infty}\theta^2 (\int(T+r)dP_\theta)\phi(\sqrt{r}(\theta-\mu))\sqrt{r}d\theta-(r\mu^2+1)$$

$$=\int(T+r)(\int_{-\infty}^{\infty}\theta^2 N(\frac{W(T)+r\mu}{T+r},\frac{1}{T+r})(d\theta))d\bar{Q}-(r\mu^2+1)$$

$$=\int(T+r)((\frac{W(T)+r\mu}{T+r})^2 + \frac{1}{T+r})d\bar{Q}-(r\mu^2+1)$$

$$=\int\frac{(W(T)+r\mu)^2}{T+r} d\bar{Q}-r\mu^2 .$$

Thus (5.2) and (5.3) yield the representation of the Bayes risk for (T, δ^*):

(5.4) $\quad \rho(T, \delta^*) = \int f\left(\frac{|W(T)+r\mu|}{\sqrt{T+r}}\right) d\bar{Q} \quad$ with $\quad f(\lambda) = \Phi(-\lambda) + c\lambda^2 - cr\mu^2$.

For $\lambda > 0$, f has a unique minimum at $\lambda(c)$ which is defined as the solution of the equation $\phi(\lambda)/2\lambda = c$. Since $|\mu|\sqrt{r} \leq \lambda(c)$ the stopping time $T^* = \inf\{t > 0 | \frac{|W(t)+r\mu|}{\sqrt{t+r}} \geq \lambda(c)\}$ satisfies $\bar{Q}\{T^* < \infty\} = 1$. Thus

$$\rho(T, \delta^*) = \int f\left(\frac{|W(T)+r\mu|}{\sqrt{T+r}}\right) d\bar{Q} \geq f(\lambda(c)) = \rho(T^*, \delta^*) \ .$$

□□□

The same type of argument can also be used for the one-sided problem. We consider the risk

(5.5) $\quad \rho(T) = \int_{-\infty}^{0} P_\theta\{T < \infty\} \phi(\sqrt{r}\theta) \sqrt{r} d\theta + c \int_{0}^{\infty} \theta^2 E_\theta T \phi(\sqrt{r}\theta) 2\sqrt{r} d\theta \ ,$

with $0 < c < 2$. For it the stopping time $T^* = \inf\{t > 0 | \frac{|W(t)|}{t+r} \geq \mu(c)\}$, $\mu(c) > 0$, is optimal. This follows from the representation of the risk

$$\rho(T) = \int e\left(\frac{|W(T)|}{\sqrt{T+r}}\right) d\tilde{Q} \quad \text{with } \tilde{Q} = \int_{0}^{\infty} P_\theta \phi(\sqrt{r}\theta) 2\sqrt{r} d\theta$$

and $e(x) = \frac{\phi(-x)}{\Phi(x)} + c(x^2 + \frac{x\phi(x)}{\Phi(x)})$, where $\mu(c)$ is the location of the minimum of e.

To explain the relation between Wald's SPRT and the RST we consider the problem of testing the sign of the drift of Brownian motion for simple hypotheses $-\theta$ versus $+\theta$ with $\theta > 0$. We take loss and costs as above (0-1 loss and costs $c\theta^2$) and restrict our considerations to a symmetric prior $G = \frac{1}{2}\delta_{-\theta} + \frac{1}{2}\delta_{+\theta}$. Since here the costs are constant it is well known from the theorem on page 197 of Wald's book that the SPRT minimizes the Bayes risk

$$\rho(T,\delta) = \frac{1}{2}(P_{-\theta}\{H_o \text{ rejected } (\delta)\} + c\theta^2 E_{-\theta}T)$$

$$+ \frac{1}{2}(P_{\theta}\{H_1 \text{ rejected } (\delta)\} + c\theta^2 E_{\theta}T).$$

Calculations similar to those as in the proof of Theorem 5.1 show that the Bayes risk can be expressed as

$$\rho(T,\delta^*) = \int g(\theta|W(T)|) d\overline{Q} \text{ with } g(x) = \frac{e^{-2x}}{1+e^{-2x}} + cx \frac{1-e^{-2x}}{1+e^{-2x}}$$

and $\overline{Q} = \frac{1}{2}P_{-\theta} + \frac{1}{2}P_{\theta}$. For $x \geq 0$, $g(x)$ has a unique minimum say at $b(c)$. Let

(5.6) $T^* = \inf\{t>0 | \theta|W(t)| \geq b(c)\}.$

Then (T^*,δ^*) minimizes the Bayes risk (5.1) with respect to the prior G.

Now we consider the testing problem for composite hypotheses $H_o: \theta<0$ versus $H_1: \theta>0$. In ignorance of the parameter $|\theta|$ we estimate it for instance by $\hat{\theta}_t = \frac{|W(t)|}{t+r}$. Then

$$\hat{\theta}_t|W(t)| = \frac{W(t)^2}{t+r},$$

which together with (5.6) shows that the RST is an adapted version of Wald's SPRT.

Finally we explain the connections between our and Chernoff's approach which uses free boundary techniques. Let $f(\lambda) = \Phi(-\lambda) + c\lambda^2$ denote the loss function which for $\mu=0$ appears in formula (5.4). The minimal posterior risk at the space-time point (x,t) is given by $\tilde{u}(x,t) = \inf E^{(x,t)} f(\frac{|W(T)|}{\sqrt{T+r}})$ where the infimum is taken over all stopping times T of the process $(W(v),v)$ starting at (x,t). Let $u(\lambda,s) = \tilde{u}(x,t)$ where $\lambda = x/\sqrt{t+r}$ and $s = \log(t+r)$. Since the infimum also includes the constant stopping time $T_t \equiv t$, it follows that $u \leq f$. By using similar arguments as Chernoff (1972) one can show that u satisfies the equations:

(5.7) $\partial_s u + \frac{1}{2}(\partial^2_\lambda u + \lambda \partial_\lambda u) = 0$

on the set $\Gamma = \{(\lambda,s) \mid u(\lambda,s) < f(\lambda)\}$,

$u = f$ on Γ^c,

$\partial_\lambda u = \partial_\lambda f$ on the boundary $\partial\Gamma$.

These equations establish a free boundary problem corresponding to the original stopping problem. It has a very simple solution since the loss function f does not depend on time. Its solution is given by $u(\lambda,s) = f(\lambda_c)$ on $\Gamma = \{(\lambda,s) \mid \lambda^2 < \lambda_c^2\}$ where $\lambda_c > 0$ denotes the location of the minimum of f on $\mathbb{R}_+$. It is given by the solution of the implicit equation $\phi(\lambda)/\lambda = 2c$.

There is an other aspect of this problem. As is well known, the optimal stopping rule for a stopping problem of a Markov-process is given by the first entrance time of the process to the set where the loss function u coincides with the largest subharmonic function below u. For the diffusion process under consideration (its generator is equal to $\Sigma = \frac{1}{2}\partial^2_\lambda + \frac{1}{2}\lambda\partial_\lambda$) this function is given by $u = f(\lambda_c)1_\Gamma + f1_{\Gamma^c}$.

The free boundary approach described above will prove as really useful for the related problem of termination of the observations after finite time. For this more complicated problem one can apply the free boundary techniques developed by Chernoff and others. Nevertheless our second viewpoint helps to see what happens in this case qualitatively. Since $\Sigma f > 0$ in all points $\lambda \neq 0$, one can conclude that the optimal stopping boundary tends to zero when one approaches the finite termination point.

References

Anscombe, F.J. (1963): Sequential medical trials, J. Amer. Statist. Assoc. 58, 365-383.

Appell, P. (1982): Sur l'equation $\partial_z^2/\partial x^2-\partial z/\partial y=0$ et la théorie du chaleur, J. Math. Pures Appl. 8, 187-216.

Armitage, P. (1975); Sequential Medical trials, Blackwell, Oxford.

Barnard, G.A. (1969): Practical application of tests with power one, Bull. ISI 43, 389-393.

Bass, R.F., Cranston, M. (1983): Brownian motion with lower class moving boundaries which grow faster than $t^{1/2}$, Ann. Probab. 11, 34-39.

Bather, J.A. (1962): Bayes procedures for deciding the sign of a normal mean, Proc. Cambridge Philos. Soc. 58, 599-620.

Borovkov, A.A. (1967): Boundary problems for random walks and large deviations in function spaces, (engl.) Theor. Prob. Appl. 12, 575-595.

Chernoff, H. (1959): Sequential design of experiments, Ann. Math. Statist. 30, 755-770.

Chernoff, H. (1961): Sequential tests for the mean of a normal distribution, Proc. Fourth Berkeley Symp. Math. Statist. 1, 79-91.

Chernoff, H. (1972): *Sequential Analysis and Optimal Design*, Regional conference series in applied mathematics of SIAM, Philadelphia.

Chernoff, H. and Petkau, A.J. (1981): Sequential medical trials involving paired data, Biometrika 68, 119-132.

Copson, E.T. (1975): *Partial Differential Equations*, Cambridge Univ. Press, Cambridge.

Cornfield, J. (1966): A Bayesian test of some classical hypotheses - with application to sequential clinical trials. J. Amer. Statist. Assoc. 61, 577-594.

Courant, R. and Hilbert, D. (1931): *Methoden der Mathematischen Physik I*, Springer-Verlag, Berlin.

Cuzick, J. (1981): Boundary crossing probabilities for stationary Gaussian processes and Brownian motion, Trans. Amer. Math. Soc. 263, 469-492.

Cuzick, J. (1981a): Boundary crossing probabilities for Brownian motion and partial sums, Preprint.

Daniels, H. (1954): Saddlepoint approximations in statistics. Ann. Math. Statist. 25, 631-650.

Daniels, H. (1974): The maximum size of a closed epidemic, Avd. Appl. Prob. 6, 607-621.

Daniels, H. (1982): Sequential tests constructed from images, Ann. Statist. 10, 394-400.

Darling, D. and Robbins, H. (1967): Iterated logarithm inequalities, Proc. Nat. Acad. Sci. U.S.A. 57, 1188-1192.

Darling, D. and Robbins, H. (1968): Some further remarks on inequalities for partial sums, Proc. Nat. Acad. Sci. U.S.A. 60, 1175-1182.

Dinges, H. (1982): Combinatorial devices for sequential analysis, Z. Wahrscheinlichkeitstheorie verw. Geb. 63, 137-146.

Doob, J.L. (1955): A probabilistic approach to the heat equation, Trans. Amer. Math. Soc. 80, 216-280.

Doob, J.L. (1984): *Classical Potential Theory and its Probabilistic Counterpart*, Grundlehren der Mathematischen Wissenschaften, 262. Springer Verlag, Berlin.

Durbin, J. (1971): Boundary crossing probabilities for the Brownian motion and Poisson processes and techniques for computing the power of the Kolmogorov-Smirnov test, J. Appl. Probab. 8, 431-453.

Durbin, J. (1985): The first passage density of a continuous Gaussian process to a general boundary, J. Appl. Prob. 22, 99-122.

Erdös, P. (1942): On the law of the iterated logarithm, Ann. Math. 43, 419-436.

Farrell, R. (1964): Asymptotic behavior of expected sample size in certain one-sided tests, Ann. Math. Statist. 35, 36-72.

Ferebee, B. (1982): The tangent approximation to one-sided Brownian exit densities, Z. Wahrscheinlichkeitstheorie verw. Gebiete 61, 309-326.

Ferebee, B. (1983): An asymptotic expansion for one-sided Brownian exit densities, Z. Wahrscheinlichkeitstheorie verw. Gebiete 63, 1-15.

Fortet, R. (1943): Les fonctions aléatoires du type de Markoff associées à certain equations linéaires aux derivées partielles du type parabolique, J. Math. Pures Appl. 22, 177-243.

Friedman, A. (1964): *Partial Differential Equations of Parabolic Type*, Prentice Hall, London.

Hartman, P. and Wintner, A. (1941): On the law of the iterated logarithm, Amer. J. Math. 63, 169-176.

Ito, K. and McKean, H.P. (1974): *Diffusion Processes*, Grundlehren der mathematischen Wissenschaften, 125. Springer-Verlag, Berlin.

Jennen, C. and Lerche, H.R. (1981): First exit densities of Brownian motion through one-sided moving boundaries, Z. Wahrscheinlich-keitstheorie verw. Gebiete 55, 133-148.

Jennen, C. and Lerche, H.R. (1982): Asymptotic densities of stopping times associated with tests of power one, Z. Wahrscheinlichkeits-theorie verw. Gebiete 61, 501-511.

Jennen, C. (1985): Second-order approximations for Brownian first exit distributions, Ann. Probab. 13, 126-144.

Khintchine, A. (1924): Über ein Satz der Wahrscheinlichkeitsrechnung, Fundamenta Math., 6, 9-20.

Khintchine, A. (1933): Asymptotische Gesetze der Wahrscheinlichkeitsrechnung, Ergebn. Math. 2, No. 4, Berlin.

Kiefer, J. and Sacks, J. (1963): Asymptotically optimum sequential inference and design, Ann. Math. Statist. 34, 705-750.

Klein, D. (1986): Absorption von Irrfahrten an krummen Rändern: Ein einheitlicher Ansatz für große und kleine Abweichungen. Dissertation, Univ. of Frankfurt.

Kolmogorov, A. (1929): Über das Gesetz des iterierten Logarithmus, Math. Annalen, 101, 126-135.

Lai, T.L., Siegmund, D. (1977): A nonlinear renewal theory with applications to sequential analysis I, Ann. Statist. 5, 946-954.

Lai, T.L., Robbins, H. and Siegmund, D. (1983): Sequential design of comparative clinical trials, *Recent Advances in Statistics, Papers in Honor of H. Chernoff on his Sixtieth Birthday*, ed. M.H. Haseb Rizvi et al., Academic Press, New York.

Lalley, S.P. (1983): Repeated likelihood ratio tests for curved exponential families, Z. Wahrscheinlichkeitstheorie verw. Gebiete 62, 293-321.

Le Cam, L. (1979): A reduction theorem for certain sequential experiments II, Ann. Statist. 7, 847-859.

Lerche, H.R. (1981): The law of the iterated logarithm for posterior distributions, "Proceedings of the Conference on Probability Theory", Brasov, Romania 1979, 347-355.

Lerche, H.R. (1982): Fluctuations of posterior distributions of quadratically differentiable experiments, Z. Wahrscheinlichkeitstheorie verw. Gebiete 59, 223-238.

Lerche, H.R. (1985): On the optimality of sequential tests with parabolic boundaries, *Proceedings of the Berkeley Conference in Honor of Jerzy Neyman and Jack Kiefer*, Vol. II, L. Le Cam, R. Olshen, eds., Wadsworth, Belmont, 298-316.

Lerche, H.R. (1986) The shape of Bayes tests of power one, to appear in Annals of Statistics.

Levy, P. (1965): *Processus Stochastiques et Mouvement Brownien*, Gauthier-Villars, Paris.

Lorden, G. (1967): Integrated risk of asymptotically Bayes sequential tests, Ann. Math. Statist. 38, 1399-1422.

Lorden, G. (1973): Open-ended tests for Koopman-Darmois families, Ann. Statist. 1, 633-643.

Lorden, G. (1977): Nearly-optimal sequential tests for finitely many parameter values, Ann. Statist. 5, 1-21.

McPherson, C.K. and Armitage, P. (1971): Repeated significance tests on accumulating data when the null hypothesis is not true, J.R. Statist. Soc. A 134, 15-26.

Neyman, J. (1969): In discussion, Bull. ISI 43, 386.

Neyman, J. (1971): Foundations of behaviouristic statistics. *Foundations of Statistical Inference*, ed. V.P. Godambe and D.A. Sprott, Holt, Rinehart and Winston, Toronto, 1-19.

Novikov, A.A. (1981): A martingale approach in problems on first crossing time of nonlinear boundaries, (engl. transl.) Proc. Steklov Int. Math. 1983, 141-163.

Petrovski, I. (1935): Zur ersten Randwertaufgabe der Wärmeleitungs- gleichung, Compositio Math. 1, 383-419.

Pollack, M. (1978): Optimality and almost optimality of mixture stopping rules, Ann. Stat. 6, 910-916.

Robbins, H. (1952): Some aspects of the sequential design of experiments, Bull. Amer. Math. Soc. 58, 527-535.

Robbins, H. (1970): Statistical methods related to law of the iterated logarithm, Ann. Math. Statist. 41, 1397-1409.

Robbins, H. and Siegmund, D. (1970): Boundary crossing probabilities for the Wiener process and partials sums, Ann. Math. Statist. 41, 1410-1429.

Robbins, H. and Siegmund, D. (1973): Statistical tests of power one and the integral representation of solutions of certain partial differential equations, Bull. Acad. Sinica 1, 93-120.

Schwarz, G. (1962): Asymptotic shapes of Bayes sequential testing regions, Ann. Math. Statist. 33, 224-236.

Shiryayev, A.N. (1978): *Optimal Stopping Rules*, Springer-Verlag, Ber- lin.

Siegmund, D. (1977): Repeated significance tests for a normal mean, Biometrika 64, 177-189.

Siegmund, D. (1985): *Sequential Analysis*, Springer-Verlag, Heidelberg.

Strassen, V. (1964): An invariance principle for the law of the iterated logarithm, Z. Wahrscheinlichkeitstheorie verw. Gebiete 3, 211- 226.

Strassen, V. (1967): Almost sure behaviour of sums of independent random variables and martingales, Proc. Fifth Berkeley Symp. Math. Statist. Probab., Univ. of Cal. Press, Vol. III, Part I, 315- 343.

Varadhan, S.R.S. (1966): Asymptotic probabilities and differential equations, Comm. Pure Appl. Math. 19, 261-286.

Varadhan, S.R.S. (1973): Strassen's version of the law of the iterated logarithm, in "Topics in Probability Theory", Lecture Notes of the Courant Institute (ed. Strook, D.W., Varadhan, S.R.S.).

Ville, J. (1939): *Étude Critique de la Notion de Collectif*, Gauthier-Villar, Paris.

Wald, A. (1947): *Sequential Analysis*, J. Wiley & Sons, New York.

Wald, A. and Wolfowitz, J. (1948): Optimum character of the sequential probability ratio test, Ann. Math. Statist. 19, 326-339.

Widder, D.V. (1944): Positive temperatures on an infinite rod, Trans. Amer. Math. Soc. 55, 85-95.

Woodroofe, M. and Takahashi, H. (1982): Asymptotic expansions for the error probabilities of some repeated significance tests, Ann. Statist. 10, 895-908.

Woodroofe, M. (1982): *Nonlinear Renewal Theory in Sequential Analysis*, Regional conference series in applied mathematics of SIAM, 39, Philadelphia.

Uchiyama, K. (1980): Brownian first exit from and sojourn over one-sided moving boundary and application , Z. Wahrscheinlichkeits-theorie verw. Gebiete 54, 75-116.

SUBJECT INDEX

Absorption of Brownian motion, 18
Anscombe's problem, 9
Appell transformation, 39

Bachelier-Levy formula, 4, 5, 42, 56
Backward induction, 101
Bayes risk, 10, 13, 100, 104, 110, 120, 121, 126, 127, 130, 132
Bayes tests of power one, 14, 100, 104
Boundary value problem, 19, 20
Brownian bridge, 38-40, 50, 55
Brownian motion as limiting model, 8

Cameron-Martin-Girsanov formula, 26
Central limit theorem, 1, 93
Clinical trial, 8, 9
Composite hypotheses, 9, 130, 133
Continuation region, 113
Cost (per observation), 9, 10, 100, 130, 132
Coverage probabilities of confidence bands, 8
Curved boundary first passage distribution, 1, 4, 17, 63, 67, 68
Cuzick's inequality, 81

Decision rule, 13, 14, 130
Diffusion equation, 18, 19, 21, 31, 39
Diffusion process, 111
Dynamic programming, 111

Entropy function, 92
Entropy (relative), 95
Equivalence of the methods of images and mixtures of likelihood
    functions, 39
Euler-Lagrange equation, 106
Exponential family, 93

First exit (passage) density, 5, 7, 77, 78
First exit (passage) probability, 6, 19, 34, 78
Formal saddlepoint approximation, 8, 92, 93
Free boundary problem, 108, 134

Girsanov transformation, 3, 26
Green's function, 20

Harmonic function of the diffusion equation, 32, 39, 40
Hazard function (rate), 8, 77, 78
Higher order approximation, 7, 67

Indifference zone, 1, 9, 10, 120, 130
Integral equation of the first exit density, 56, 57, 58, 80

Kaplan-Meier estimator, 8
Kolmogorov-Petrovski-Erdös test, 1, 4, 77-79, 86

Lecture Notes in Statistics

Vol. 26: Robust and Nonlinear Time Series Analysis. Proceedings, 1983. Edited by J. Franke, W. Härdle and D. Martin. IX, 286 pages. 1984.

Vol. 27: A. Janssen, H. Milbrodt, H. Strasser, Infinitely Divisible Statistical Experiments. VI, 163 pages. 1985.

Vol. 28: S. Amari, Differential-Geometrical Methods in Statistics. V, 290 pages. 1985.

Vol. 29: Statistics in Ornithology. Edited by B.J.T. Morgan and P.M. North. XXV, 418 pages. 1985.

Vol. 30: J. Grandell, Stochastic Models of Air Pollutant Concentration. V, 110 pages. 1985.

Vol. 31: J. Pfanzagl, Asymptotic Expansions for General Statistical Models. VII, 505 pages. 1985.

Vol. 32: Generalized Linear Models. Proceedings, 1985. Edited by R. Gilchrist, B. Francis and J. Whittaker. VI, 178 pages. 1985.

Vol. 33: M. Csörgő, S. Csörgő, L. Horváth, An Asymptotic Theory for Empirical Reliability and Concentration Processes. V, 171 pages. 1986.

Vol. 34: D.E. Critchlow, Metric Methods for Analyzing Partially Ranked Data. X, 216 pages. 1985.

Vol. 35: Linear Statistical Inference. Proceedings, 1984. Edited by T. Caliński and W. Klonecki. VI, 318 pages. 1985.

Vol. 40: H.R. Lerche, Boundary Crossing of Brownian Motion. V, 142 pages. 1986.